The Teeth of the Lion

Frontispiece. This composite image of the major anatomical elements of the dandelion, including those of the different reproductive stages, appeared in volume 1 of F. E. Köhler, *Köhler's Medizinal-Pflanzen in naturgetreuen Abbildungen mit kurz erläuterndem Texte . . .*, published in Gera, Germany, in 1883 (1887). (© 1995–2006 by, and used with permission of, the Missouri Botanical Garden)

The Teeth of the Lion

The Story of the Beloved and Despised Dandelion

by
Anita Sanchez

illustrations by
Joan Jobson

The McDonald & Woodward Publishing Company
Blacksburg, Virginia

The McDonald & Woodward Publishing Company
Blacksburg, Virginia, and Granville, Ohio

The Teeth of the Lion: The Story of the Beloved and Despised Dandelion

Text © 2006 by Ana Maria Sanchez
Figures © 2006 by Joan Jobson

All rights reserved. First printing October 2006
Printed in the United States of America by
McNaughton & Gunn, Inc., Saline, MI

10 9 8 7 6 5 4 3 2 1
16 15 14 13 12 11 10 09 08 07 06

Library of Congress Cataloging-in-Publication Data

Sanchez, Anita, 1956-
  The teeth of the lion : the story of the beloved and despised
        dandelion / Anita Sanchez.
    p. cm.
  Includes bibliographical references and index.
  ISBN 0-939923-22-X (pbk. : alk. paper)
  1. Common dandelion. I. Title.
QK495.C74S25 2006
583'.99—dc22

2006027672

iv

Contents

To George

Unless . . . .

I have great faith in a seed. Convince me that you have a seed there, and I am prepared to expect wonders.

— Henry David Thoreau

Acknowledgements

I've always dreamed of writing a book, but I never realized how many people it takes to write one—as many as the seeds of the dandelion, it seems. I could go on thanking these friends and colleagues forever, but I'll try to keep it short; it's impossible to express how much I owe them.

I would like to offer thanks to Dr. Harvey Alexander of the College of St. Rose in Albany, New York, who read the manuscript and gave me many valuable ideas and corrections. Thanks to Barbara Friedman, DDS and medical historian, who contributed many helpful insights on medicine, and introduced me to her friend Dr. Culpeper. Grateful thanks to Dr. Ward B. Stone, Wildlife Pathologist of the New York State Department of Environmental Conservation, who read the material on pesticides. Any errors that remain are mine alone.

Many thanks to those friends who read the manuscript and encouraged me to keep on going: Gage Evans and Marla Zylstra. Thanks also to the courteous and endlessly patient staff of the Bethlehem Public Library in Delmar, New York.

Thanks to my wonderful colleagues at the New York State Department of Environmental Conservation, especially those at Five Rivers, the best place to work in the world.

Thanks to my partner, Joan Jobson, for the beauty of her art, and her grace under pressure: always quick on the draw!

And thanks to my family: to my mother, who taught me to love books — my earliest memory is of her reading *Robin Hood* to me — and to Gramps, who taught both of us that there's nothing better than a good story.

Thanks to my husband, George, for help in a million ways — and for always encouraging me to follow my dreams, and giving me the elbow room to do it.

Thanks to Timothy Steele, for sharing his ideas and insights on nature, and for helping me make dandelion pie.

Thanks to Alex Steele, who has contributed greatly to the making of this book. During many late-night editing sessions, his sharp eyes detected errors, and his analytical questions made me think, and re-think. He refused to tolerate inaccuracies, vagueness, extraneous adverbs, or anything remotely "cutesy," and he ruthlessly insisted that I write as good a book as possible. A skilled writer, he is the author of the book's first two paragraphs.

Introduction

Early in the seventeenth century, two armies sailed from Europe, bound for North America. They landed at the same time and place, and encountered the same conditions. One army struggled for survival, suffering an enormous casualty rate. The other invaders quickly established a foothold on the coast and within a few years had spread far across the continent, increasing their numbers at an exponential rate.

Humans made up the first group of invaders — the ones that almost died off. The details of the other invasion aren't in the history books, but it really happened. The wildly successful invaders were plants — the sometimes beloved and often despised lawn plants we call dandelions.

Dandelions. Whether you love them or whether you hate them, dandelions are, perhaps, the most familiar plant in the world. They're the one species of plant that just about anyone can identify at a glance, as familiar to humans as the dog.

This most common of plants is now despised as a weed by many, but it wasn't always that way. Dandelions were once a valued commodity, purposely transported by humans across oceans and continents. Americans today spend forty billion dollars annually on lawn care, and a hefty part of that budget goes to the attempt to eradicate dandelions — the very plant that was brought to this country by its earliest European settlers, who prized the plant for its medicinal powers and nurtured the cheerful golden flowers for their beauty. Not too long ago, prize-winning dandelions were exhibited at county fairs — one variety was patriotically christened the "American Improved."[1] Gardeners used to weed out the grass to make room for the dandelions.

What happened? In an amazingly short space of time — less than one human life-span — this loved garden flower became the most unpopular plant in the neighborhood. Thirty million acres of the United States are now lawns, and an estimated eighty million pounds of pesticides are used on them annually.[2] Probably no other plant in the world undergoes such a barrage of deadly chemicals; humans have attempted to exterminate dandelions with a passion that's usually reserved for cockroaches or tarantulas. Yet the dandelion remains.

How does the dandelion do it? What's the secret of its success? In these days of environmental destruction, thousands of species are on the verge of extinction. Not just California condors and blue whales. Plants are in peril, too; wild orchids and ferns, painted trillium and pink lady-slippers: whole species are declining in spite of frantic efforts to preserve them. An estimated 100,000 species of plants are in danger of extinction worldwide.[3] It might be worth our while to examine how it happens that the dandelion can thrive, sometimes in the most inhospitable of habitats, defying humans' best efforts to get rid of it.

Dandelions are fast growers, the sunny yellow flowers going from bud to seed in a matter of days. But they are also long-lived — an individual plant can live for years, so that the dandelion lurking in a corner of the playground might be older than the children running past it. Dandelion plants have long roots, which sink deeper and deeper into the soil through the seasons. Dandelions have sunk their roots deep into history as well.

I have prowled many a library, searched the dusty shelves of bookstores, and surfed the Internet in search of information about dandelions. Finding information wasn't a difficult task. There's so much material on dandelions that I was often in danger of drowning rather than surfing. I discovered tips for growing bigger and better dandelions, and sure-fire methods for exterminating them. I read of myth and fact, lore, magic spells and love charms; of dye made from dandelion roots and rubber made from dandelion sap; recipes for such delicacies as dandelion chicken Alfredo, dandelion ice cream, and dandelion wine; and some of the world's most ancient prescriptions recommending dandelions as infallible remedies.

I think it's fair to say that no other plant in the world has been linked with such a wildly diverse assortment of cultures and times. Early American Shakers advertised Extract of Dandelion as a liver tonic; Japanese gardeners formed dandelion horticultural societies to celebrate the beauty of the golden blossoms. Arab and Chinese physicians wrote of its medical benefits a thousand years ago, and herbalists swear by it to this day. Greek philosophers, ancient Jewish rabbis, Native American shamans, New England witches, . . . to each of these, the dandelion showed a different face, gave a different message. Children, gardeners, botanists, exterminators; Pilgrims, Persians, Egyptians, twelfth-century Benedictine monks — all have bent over this little roadside weed and considered it well, pondering how it could fill their own particular need.

Let us join them.

Notes to Introduction

1. Sturtevant, E. Lewis. *Edible Plants of the World.* New York, NY: Dover Publications, 1972. In 1871, no fewer than four varieties of dandelion were exhibited at the Massachusetts Horticultural Society, under the names

French Large-leaved, French Thick-leaved, Red-seeded, and the proudly-named American Improved Dandelion.

2. Wargo, John. *Risks from Lawn Care Pesticides*. North Haven, CT: Environment and Human Health, Inc., 2003. Environment and Human Health, Inc. (EHHI) is a non-profit organization made up of doctors, public health officials, and policy experts dedicated to the reduction of environmental health risks to humans. See more at *http://www.ehhi.org/reports/lcpesticides/summary.htm*.

3. This estimate of threatened species is that of Botanic Gardens Conservation International (BGCI), a professional organization of botanic gardens across the world, dedicated to the study and preservation of rare plant species. See more at *http://www.bcgi.org*.

Chapter 1

Putting Down Roots

Massachusetts
About 1621

A woman kneels in the rich, dark soil of a garden. Golden sunlight pours down on the freshly-turned earth, and the comforting warmth helps her to forget the cold. After a winter filled with hardship, death, and sorrow, it had seemed as if the sun would never shine again. But now the grass is greening, and spring has come at last.

She crumbles the soil between her fingers, preparing it carefully for the seeds. It is vitally important that these seeds grow: she is planting an herb garden, of plants carefully chosen for their healing powers. She lifts her tired eyes and scans the forbidding tangle of trees crowding close to the huts of the little settlement. Here there are no doctors' offices or apothecary shops, only wilderness. The plants she grows in her garden will be medicines that might make the difference between life and death for her family in this terrifying New World.

She etches a furrow in the fertile soil, then carefully sprinkles in a row of tiny brown specks. She pats the soil over them lovingly, murmuring a prayer over the precious seeds: a heartfelt prayer that dandelions will grow, where before there was only the barren grass.

~

The overcrowded little vessels beat their way across the icy waters of the North Atlantic, and the storms were fierce. The passengers were seasick and miserable, cramped below decks in the damp, smelly hold, and as they tossed on the rough seas the voyagers tried desperately to keep their children and their possessions safe and dry — especially what was perhaps the most important possession of all: seeds. These travelers were headed for a land filled with unplowed and fertile soil — a whole New World — and they were planning to put down roots.

They were not the first Europeans to invade North America, by any means. But most of the previous comers — the fishermen and fur trappers, and the gold-hungry settlers of Jamestown — were just passing through, grabbing whatever resources the new land would yield, and heading for home to cash in as quickly as possible.[1] In 1620, with the sailing of the *Mayflower,* a new type of invasion began along the beachheads of the New England coast. Whole families embarked for the New World, and they intended to stay; they packed up the kids, and the dogs and chickens; they brought tables and chairs, pots and pans, blankets and cradles. And they made sure to pack the tools of farming: shovels, hoes, and axes.

These travelers, so far separated from us in time, came from an environment that in many ways was similar to ours: an ordered world of houses and streets. Many of them came from towns or cities, places where you could run to the bakery to buy a loaf of bread, or go to the theater to see a play; you could call a constable for help if there was trouble, or stop by the doctor's office to pick up medicine if your kid happened to be sick. Their world was constructed, organized; it was a human-dominated environment, a controlled landscape, with the arable fields carved into small postage-stamp plots that had been plowed for generations. The wilderness had long ago been beaten back, pushed out of sight and all but forgotten.[2]

But the dim green coastline they were approaching presented a very different scene from the skyline of church spires and tall buildings they had sailed from. The passengers on the first ships to approach the New England coast must have turned to each other with blank faces and frightened hearts when they saw what confronted them: a wilderness of trees. In those days, it's said, a squirrel could travel tree to tree, from Cape Cod to the Great Lakes, and never need to set a paw to the ground; east of the Mississippi, most of the continent was one big forest.

The word "wilderness" has a pleasant ring to my twenty-first-century ears; the idea of heading off into a wilderness sounds enjoyably adventurous when it's enclosed within the boundaries of a national park. But in other centuries, "wilderness" was a noun generally preceded by adjectives like "hideous" and "desolate."[3] The first European settlers saw the forest as a dismal place, full of "cruell, barbarous, and most trecherous Indians" skulking in the shadows.[4] North America was full of the promise of wealth, but its wildness was an obstacle to be overcome; this wild new land was not a congenial place for humans, or a good habitat for the plants they needed to grow in order to survive — plants like dandelions.

◇

How did dandelions, a Eurasian plant known to botanists as *Taraxacum officinale*, get to the New World? No one knows for sure. Theirs was a quiet invasion. Legend has it that the Pilgrims first set foot on Plymouth Rock, and the big gray boulder is a monument today, enshrined under an imposing imitation of a Greek temple. But no stone marks the spot where dandelions first came ashore.

The ships that crossed to the New World were often filled with rocks and earth for ballast, which was then dumped dockside. The piles of dirt contained seeds, dandelion seeds almost surely among them, and soon the little yellow flowers probably were poking out of the sandy soil near bays and rivers. Other dandelion seeds undoubtedly infiltrated the new environment in more subtle ways, arriving in pantcuffs, rolled up in socks, or stuck in a cracked boot sole.

But dandelions came not only as stowaways. They traveled first-class as well as steerage, having been invited on board as welcomed passengers, so to speak. Dandelions were one of the very first Eurasian plants to invade the Western Hemisphere—and some arrived in carefully wrapped packages of garden seeds.

～

Why on earth would anyone plant dandelions?

Well, you can eat dandelion greens, of course. But seventeenth-century folk didn't favor leafy vegetables much, being mostly no-nonsense meat-and-bread eaters, although in springtime they occasionally enjoyed a mess of "sallet herbes."[5] And the stern Puritans were not much given to indulging in dandelion wine. Dandelions were neither food nor drink — they were medicine.

While the men did the heavy work of cutting down the trees and plowing, gardening was the women's chore. In the newly-cleared sunny spaces, they set themselves to create herb gardens, little square-edged plots seeded with familiar, homelike plants — small and hopeful patches of order in the chaos of the wilderness. The herb gardens were filled with plants that were the housewives' indispensable medicine chest. Some plants were potentially life-saving remedies, others were cures for the little ailments of life, the botanical equivalents of Pepto-Bismol, Kaopectate, aspirin, sleeping pills, and Band-Aids.

All parts of the dandelion plant — leaf, root, and flower — had been known for millennia as efficacious remedies against a host of ailments, and were hailed in popular "herbals," or books of plant lore, as powerful medicine.[6] The anxious wives and mothers, facing so many hardships in the New World, eagerly planted the seeds of such useful plants. Although no one understood why dandelions were good medicine, everyone knew that they worked. Dandelions' golden blossoms were considered one of the most useful flowers in the garden.

There's an old saying about how to choose the plants for your garden: "Some for use and some for delight." The versatile dandelion brought not only healing, but joy.

The early settlers planted dandelions for the same reason that modern-day travelers take refuge at a McDonald's or a Holiday Inn: when you're far from home, the greatest comfort lies in the familiar. The American wilderness was terrifying — often lethal — *terra incognita.* So, almost from the hour they landed, the European settlers began the task of making the New World as much as possible like the Old World they had fled.

To the bewildered children who had been uprooted from their old life, everything must have seemed alien. Even the very flowers in their dooryards were unfamiliar, and the homesick youngsters must have missed daisies and buttercups, red and white clover, Queen Anne's lace, and dandelions — none of which were found in North America till humans brought them. The cheerful face of the dandelion provided a sweet reminder of home.

When a child blows on a dandelion puffball, the gray globe shatters, and the seeds fly high. Depending on the wind, the little gossamer parachutes can travel a few feet or hundreds of miles. A single dandelion flower head could produce two hundred seeds, and a germination rate of ninety percent is not unusual. The dandelions were on the move.

Soon there was no need for anyone in the New World to plant dandelions. No garden fence was ever built that could contain the aggressive little plants, and they began elbowing their way into fields, roadsides, and pastures. Within a few years of their arrival, dandelions were solidly entrenched in their new empire. The human immigrants spent long, exhausting years of toil to build a life in the wilderness, but the dandelions made themselves right at home. A botanical survey of New England in 1672 reported them as well-established plants.[7] Spaniards also brought dandelions to the New World as medical herbs, importing them into California and Mexico; the French introduced them into the wilds of Canada. Soon the little rosettes of jagged-edged leaves were spreading from the edges of the continent towards the heartland, springing up behind the settlers like green-and-yellow footprints.

It's interesting to speculate on the thoughts of the continent's original inhabitants as they began to notice the small invader moving inland from the coast. They quickly learned of its healing powers; Native American nations across the land, from the Algonquians of eastern Canada to the Aleuts of Alaska, adopted the dandelion into their pharmacopoeia as soon as it made an appearance in their neck of the woods.[8]

But it isn't recorded if any of the Wampanoag or Algonquian, Iroquois or Cherokee felt a sense of unease at the persistence of this flower that sprang up like magic as soon as the newcomers cut a tree or plowed a meadow. There's no way to know if the First Peoples recognized in the cheery little dandelion an omen of changes to come, a warning that everything in their world was about to shatter, like a dandelion puffball in a hurricane wind.

Notes to Chapter 1

1. Kurlansky, Mark. *Cod: A Biography of the Fish That Changed the World.* New York, NY: Penguin Putnam, 1997. European fishermen had likely been visiting the New England coast, well known to be a fabulously rich fishing ground, for centuries before permanent European settlements were established in the region. Kurlansky suggests that Basque fishermen had been fishing off the New England coast long before Columbus "discovered" the New World.

2. Willison, George F. *Saints and Strangers; Being the Lives of the Pilgrim Fathers, & Their Families, With Their Friends and Foes.* Orleans, MA: Parnassus Imprints, Inc., 1945. While most of the passengers on the *Mayflower* came originally from small farms and villages in England, some of them had been living for years in the pleasant Dutch city of Leyden, a thriving commercial center of some 80,000 people.

3. Bradford, William. *Of Plymouth Plantation.* New York, NY: Random House, 1981. First published in 1856 as *History of Plymouth Plantation.* William Bradford was one of the leaders of the "Saints," as the Pilgrims called themselves, and a passenger on the *Mayflower* in 1620.

4. *Ibid.*

5. Travers, Carolyn Freeman, ed. *The Thanksgiving Primer.* Plymouth, MA: Plimouth Plantation, Inc., 1987.

6. John Gerard's classic, *The Herball; or General Historie of Plantes,* written in 1597, and William Turner's *A New Herball,* published in three parts in 1551, 1562, and 1568, were two well-known books that praised the dandelion's medicinal benefits.

7. Josselyn, John. *New-England's Rarities Discovered.* Boston, MA: Massachusetts Historical Society, 1972. First published in 1672. Josselyn made a methodical survey of the plants of New England, distinguishing native plants from "plants as have sprung up since the English Planted and kept Cattle in New-England."

8. *http://www.ars-grin.gov,* a website of the US Department of Agriculture's Agricultural Research Service.

Chapter 2

The Golden Road

Nebraska
1843

A woman clings to the wooden seat as the covered wagon bounces along the rough track. The trail runs, straight as an arrow, over the low hills of the prairie. Oxen toil to pull the wagon, heavily laden with all the necessities of pioneering: hammers and nails, quilts and frying pans, axes, plows and seeds. She gazes over the endless sea of grass. There are no streets or houses to see, no gardens or crop fields here in the wilderness. She is far from home.

The silence is as immense as the prairie, except for the whisper of the hot, dry wind through the grass blades, and the monotonous creak of the wagon wheels. Her skin is burned by the sun, and her lined face is weary, but her eyes are calm and untroubled. She is fortified with the certainty that she and her family are on the right path, the path ordained for them. They can see the way clearly: the road to the Promised Land stretches bright before them. The lonely trail over the prairie is lined with golden flowers.

~

The white wagon-tops of the prairie schooners skimmed like sails across the sea of grass that was once the vast North American prairie. Ships crossing the ocean leave no permanent trace behind, but when those first covered wagons cruised along, they left the tracks of their wheels, a long, narrow scar stretching across the land. The next wagons followed in those tracks, further splitting and breaking the thick prairie sod.

The roots of the billions of grass plants that covered the prairie were interwoven in a dense web that was almost impossible to sink a shovel though. The sod was like armor — all but impervious to blasting winds, torrential rains, or the hoofs of grazing bison. Animals chewed the tops of the grasses, but fresh growth sprang from the vast network of roots which held the soil in place like outstretched hands.

But as wagon followed wagon, the breaks in the sod began to widen; they became ruts, then trails. The protecting armor of sod was breached, leaving the soil exposed. Seeds were blown into the wheel ruts, and flowers took root: sunflowers, coneflowers, black-eyed Susans, and dandelions. The stripes of yellow flowers over the prairie became so apparent that pioneers told of the golden road that God had sent to lead them westward.

Of course, the sod had sustained wounds before those first wheel marks. Nature has an endless supply of disasters up her sleeve. Lightning leaps down from the sky and starts a fire, a river floods and scours its banks, earthquake, landslide: inevitably, the earth's skin is from time to time torn open, the protecting layer of vegetation peeled away. And when that happens, the land is left vulnerable to the worst disaster of all — erosion. Once the soil is exposed, it begins to vanish; wind and water steal it, speck by speck, till what remains is a bare, hard surface, a wound on the earth where it seems that nothing can grow.

After an injury, wise mothers apply Band-Aids. Nature's Band-Aids are the tough, aggressive plants called

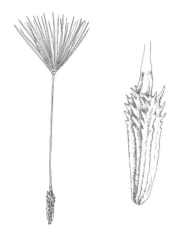

Figure 1. A dandelion seed, with parachute attached, at left, and the achene, enlarged, at right.

seral species. These are hardy plants that thrive in the conditions that occur after disaster, springing up on soil that seems as desolate as a war zone.

Dandelions are a seral species. They move into a new habitat quickly, like medics rushing onto a battlefield. Each dandelion seed (Figure 1) is a dry, hard, brown speck an eighth of an inch long, known in botanical terms as an achene, with minuscule barbs arranged along each edge of the seed. Hanging under its parachute, the pointed seed gently comes to earth and touches down like a practiced paratrooper, feet first. The umbrella-shaped parachute remains erect, spread protectively overhead, and each touch of breeze makes the seed tilt back and forth, embedding itself more firmly with every movement. The seed slowly penetrates the soil, working its way in deeper, like a barbed arrow.

At the first touch of moisture, dandelion seeds germinate with explosive speed (Figure 2). Growth begins at the

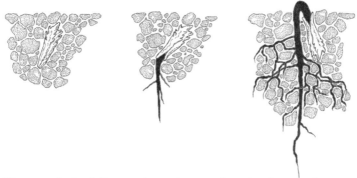

Figure 2. A dandelion seed germinates, thrusting its growing root system (shown here in black) between soil particles as it enlarges and extends its reach. The above-ground portion of the plant begins to emerge at right.

soil surface or near it, when the temperature is at least 50° F, though light and warmth increase the rate of germination, which is at the max at about 77° F. Tiny rootlets snake their way between grains of soil, in search of water and nutrients which flow back up the lengthening roots to power the growing plant (Figure 3). The web of dandelion roots pry the close-packed soil particles apart, effectively roto-tilling the earth,

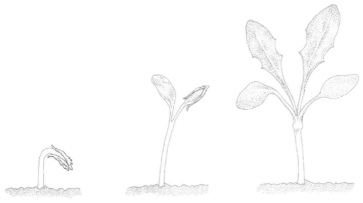

Figure 3. A dandelion seedling emerges from the ground.

while at the same time they hold the loosened soil in place. A long taproot sucks calcium and other minerals from deep in the ground, and carries the nutrients upwards to the leaves. Soon the first layer of dandelion leaves decomposes into a nutrient-filled compost. Dandelions enrich the soil for other plants; they fertilize the grass.

Like the determined folk in the covered wagons, seral species are pioneers; they blaze a trail which others, less hardy, can follow. And like all pioneers, they fundamentally change the nature of the place they come to inhabit. The newly fertilized and loosened soil is colonized by other plants: grasses are usually the next to move in, then shrubs and trees. Inevitably, the sun-loving seral plants fade away in the shadows. The dandelion prepares the way for its own extermination.

That's how nature heals natural disasters. The seral species, the dauntless pioneers, are drawn to the ecological land of opportunity; they move in and then are replaced by other species until the armor of green is restored, in the age-old natural process known as succession.

But a wheel rut is not a natural disaster. It's a human-made disaster. As soon as the seral species move in to start the healing work, another wheel rolls by. The process starts all over again. But then another wheel goes by. It's a state of permanent catastrophe.

The settlers moved west in the prairie schooners, and were inexorably followed by railroads, and then automobiles. The prairie was relentlessly plowed until only the tiniest postage-stamp scraps remained. Prairie soil was rich, dark, free of rocks, a dozen feet deep in places. To the Irish, German, or Polish farmers, coming from lands of exhausted soil that had been farmed for centuries, the black prairie dirt was fertile beyond belief. To the boys from Massachusetts or New

Hampshire, accustomed to grubbing rocks and stumps out of the hardscrabble hills of New England, the flat, treeless prairie must have seemed like heaven.

The first plowing of virgin prairie — "sod-busting" — was brutally hard work, but technology soon solved the problem. Steel-bladed plows snapped the tough roots of prairie grasses with reports like pistol-shots. Big bluestem and bunchgrass, as well as orchids, wild roses, and cone-flowers, were replaced by other, more lucrative species of plants. The buffalo-grazing grounds of Kansas, Oklahoma, and the Dakotas were turned into one big wheat field. With ruthlessness and lethal efficiency, the European settlers displaced the Native Americans, European cows displaced bison, and European plants displaced the native grass and flowers.

The prairie had remained essentially undisturbed for millennia. But the invasion of the plow was an ecological catastrophe. The protecting network of roots was torn apart, but few farmers foresaw the deadly erosion that would soon follow. The starvation and misery of the Dust Bowl were implicit in the first steel plow blade that opened up the sod.[1]

⌇

Dandelions came west in the prairie schooners — brought on purpose, mind you, and planted as garden flowers.[2] The pioneer women were as lonesome and homesick as the Pilgrim women had been, two hundred years earlier. Like the forests of New England, the prairie seemed a frightening, barren, Indian-ridden wilderness, and the cheerful yellow flowers clustered in the dooryard were like a neighbor's friendly face.

Dandelions were still a perennial favorite medicine. The Shakers sold dandelion products for decades, and cultivated dandelions in their acres devoted to medicinal herbs. Shaker products were widely known to be of high quality and were popular all over the eastern United States — the Shakers even got a request from the famous Covent Garden Market in London, asking the price for shipping dandelions (irony of ironies!) back over the Atlantic to England.[3]

And dandelions were growing in popularity as a food. Horticulturists developed many varieties of dandelion greens, and came up with tasty choices for the home gardener; the French Large-leaved Dandelion, for example, was considered a culinary delight. Dandelion greens had long been a European treat, sold in the food markets of Paris, and served at famous restaurants. But it was not only wealthy gourmets who enjoyed their sharp flavor. One Ninevah Ford, a covered-wagon pioneer on the Oregon Trail, is recorded as bringing dandelion seed with him so that he could be sure of having plenty of his favorite dandelion salad. Ninevah fathered eleven children, survived two wars, was elected to Congress, and lived to be eighty-two, perhaps a testimonial to the efficacy of his favorite food.[4]

Soon, of course, there was no need for Ninevah to worry about a lack of dandelions; the plants jumped the garden fences as they always have, and found themselves in a very congenial habitat. The plowed-up prairie was full of sun and disturbed soil, just the sort of conditions where dandelions thrive.

The descendants of those first invaders flourish in the crop fields of the Midwest to this day. But dandelions aren't a significant agricultural weed; a farmer hardly notices the little plants as the tractor tears up the soil. A six-foot-tall cornstalk has no trouble at all shading out a six-inch-tall dandelion, and no one cares if there's a few dandelions lining a corn field. No, it's not farmers who curse dandelions as being the Devil's invention.

~

As the prairie was cleared, and the land became "civilized," the landowners began to relax and enjoy themselves a little. The second and third generations of settlers built comfortable houses instead of sod huts and log cabins. And early in the twentieth century, homeowners began to pay more attention to the plants in front of their houses. Backyards had long contained gardens of nutritious or useful plants, but now, urged by garden clubs and civic organizations to beautify America, people across the country began to plant a strange thing in their front yards: not flowers, not vegetables, but grass.

This grass was not intended for any of the age-old uses of grass: it wasn't planted to feed cows or horses, provide fuel, or stuff mattresses; it was purely for decoration. This new use of grass was a foreign tradition, a status symbol of wealthy European landowners who had enough acreage that not every inch need be devoted to raising food, and enough servants to keep the grass cut short. One can imagine the original sod-busters scratching their heads over this strange hobby of their grandchildren: raising grass that was carefully trimmed — not plowed or harvested, but mowed many times a season — to become a lawn.

A lawn is, of course, a pleasant place. The very word "lawn" has a casual, lazy ring to it, bringing to mind hammocks and iced tea, softball games and summer vacations. A lawn, to most humans, is a congenial spot to sit down and rest for a while. But there's a proverb that says that if it's God who invented grass, it's the Devil who invented lawns.

These days, a lawn seems like the most natural thing in the world, a green stretch of nature among the city streets and suburban housing lots. But there's nothing less wild than a lawn. It's a crop-field just as much as is a field of wheat or soybeans, and as quick to disappear without constant human maintenance. A lawn is continually being altered by humans; ecologically speaking, it's a permanent catastrophe, exactly like the wheel ruts on the prairie. The native protecting vegetation has been stripped aside, the natural rhythm of succession altered. Healing can never take place. Long before the grass can grow tall to shade out the seral species, the mower roars across the landscape, cutting plants back to the ground. As far as nature can tell, a lawn is a perennial war zone.

Every time you mow the lawn, putting the brakes on succession, you send a clear ecological message — a disaster has occurred! And how does nature react to catastrophe? What plants inevitably arrive, like medics, to heal the wounded earth?

The pattern repeats as it has for countless eons. Right after a disaster, nature sends in the Marines, so to speak. The hardy pioneers invade the environment that they're best suited to exploit. Humans have created an artificial habitat that is utterly ideal for seral species — like dandelions.

Notes to Chapter 2

1. Leopold, Aldo. *A Sand County Almanac*, New York, NY: Oxford University Press, 1968. First published in 1949. Born in 1887, Aldo Leopold saw first-hand some of the transformation of wild prairie to farmland. Leopold's classic work helped make people aware of the destruction of the prairies and proposed a land conservation ethic that we are still struggling to reach today.

2. This was, however, not the first time that dandelions were planted in gardens for their beauty. In Japan, for example, whole horticultural societies had been created to celebrate the beauty of dandelions and create colorful hybrid flowers for the garden. The Japanese came up with several varieties, including white, orange, black, and copper-colored dandelions.

3. Burns, Deborah. *Shaker Cities of Peace, Love and Union.* Hanover, NH: University Press of New England, 1992.

4. *http://www.oregonpioneers.com/1843/htm,* a website with much fascinating information gathered by Stephenie Flora and others on The Oregon Territory and its Pioneers, and a recipient of the 1998 American Local History Network Award. The information on Ninevah Ford is from a section of the website titled *Emigrants to Oregon in 1843.*

Chapter 3

Early Seed

Concord, Massachusetts
1861

A tall, bearded man crosses a summer meadow with a long-legged stride. He knows the way well; he has traveled a good deal in the woods and fields surrounding this, his home town of Concord. He knows where to pick the sweetest berries, where the wild orchids grow, where foxes and woodchucks have their dens.

He spots something of interest, stops, and bends down. Then he lies in the sun-warmed grass, studying a small plant specimen with a magnifying glass. A farmer in a nearby field leans on his hoe and watches in disbelief — a grown man, flat on his stomach, staring at weeds!

Finally the bearded man gets to his feet, holding a dandelion carefully between thumb and forefinger. He takes off his straw hat, places the plant carefully inside, and claps the hat back on his head. He nods to the farmer, and saunters off, whistling.

The farmer shakes his head in amusement, and disgust, and goes back to his hoeing. Just another proof of what the neighbors have been saying for years: Henry David Thoreau is just plain crazy.

~

Glance down at a dandelion clawing its way up through a crack in the pavement and, like Thoreau, you have to take your hat off to it. When the great American writer and naturalist Henry David Thoreau found a plant specimen that interested him, he would often tuck the plant inside his hat, in order to carry it home for later study.[1] It's the best way I've found to transport fragile specimens — Thoreau even built a tiny scaffolding inside the hat to lay the plants on — and moisture from your head keeps the leaves nice and fresh. Of course, you may get a bug or two in your hair. But Thoreau wouldn't have minded that.

He was a true naturalist, intrigued by everything he encountered, every daily, trivial manifestation of nature: insects, briers, mud puddles, squirrels. Where his industrious Concord neighbors saw only weeds and varmints, Thoreau saw wonder and beauty — even in a yard filled with dandelions.

After his daily rambles, Thoreau would study his specimens by lamplight, making detailed notes and sketches of the humblest roadside plants. Always both poet and scientist, his descriptions of thistles, acorns, brambles, and dandelions are filled with words of beauty. But he was a meticulous observer. Thoreau painstakingly recorded the annual blooming dates of hundreds of species of plants in the Concord area, sometimes visiting the same shrub or wildflower day after day, year after year, in order to note the precise date on which the first blossom opened.

The year 1861 was one of turmoil in Thoreau's quiet home town, as it was everywhere in America. The long-smoldering controversy over the abolition of slavery had built to the inevitable explosion, and the nation was at war. Peaceful summer meadows would soon be blasted with cannon fire and strewn with mangled bodies. Thoreau had been an avid abolitionist, but as the mind-numbing body count began to fill the newspapers, he increasingly turned his attention to the tranquil world of plants. As though to counter the coming harvest of death, Thoreau focused his attention on a study of life; he embarked on a project that he expected might take a decade, a rigorous and detailed examination of nature at its most basic beginnings — the dispersal of seeds.

Thoreau filled thousands of pages of notebooks with descriptions of natural history. And he devoted a few of his words to the wonders of the humble dandelion. He marveled at its precise structure, and admired the adaptations that enable it to survive so deftly. Above all, he was impressed by the remarkable efficiency with which the plant spreads its seeds.

"I have great faith in a seed," he remarked. "Convince me that you have a seed there, and I am prepared to expect wonders."[2] Thoreau, who spent so much time in dandelion territory that he often must have had to brush the seeds off his jacket and pick them from his beard, really understood the matter. Dandelions can sprout in places that seem little short of miraculous, barren habitats where almost any other plant would throw in the towel.

Dandelions seem to thrive along the edge of highways, sinking roots into rock-hard soil that's driven over by cars, parked on by eighteen-wheeler trucks, and scraped and salted by giant municipal snowplows in wintertime. The tender green leaves shove their way through gravel and slice through layers of blacktop. They're found world-wide, spread across the planet on every continent except Antarctica, below sea level and above tree-line, growing in fertile fields and desert canyons, lining the clay soil of riverbanks and basking in the sun on sand dunes.[3]

And, unlike most plants that have a history of garden use, dandelions don't seem to need any help from human hands. Thoreau dabbled in farming while he was living at Walden Pond, and knew well how many hours you have to spend hoeing beans. I think of all the time I've spent on my knees, coaxing roses or tomatoes to grow in carefully prepared and fertilized soil. It's hard to restrain a sigh of exasperation at the dandelion flourishing in the driveway gravel.

It's not that dandelions *prefer* poor soil. If you plant a dandelion seed in rich black compost, and then weed out the competing plants and throw in some fertilizer, you'd better stand back. A dandelion grows with amazing rapidity, and can sprout a waist-high blossom with a long neck like a giraffe, and a three-inch flower head on a thick stalk. No, dandelions don't mind at all when life is a bed of roses. But when life hands the dandelion a challenge, the dandelion rolls up its sleeves (metaphorically speaking) and gives it a try. And they like to get an early start.

"About the 9th of May," Thoreau noted one spring morning, in his sprawling handwriting, "we begin to see the dandelion already gone to seed here and there in the green grass of some more sheltered and moist bank, when we are looking for the earliest flowers . . . that little seedy spherical system which boys blow . . ."[4]

The lanky naturalist wandering the fields was, at this point in his life, a profoundly solitary man. Although he was barely more than forty years old, his portrait shows a man with slumped shoulders, a lined face, and tired eyes. He had no wife, no children. He may have watched the youngsters in the fields a little wistfully, as they played in the meadows blowing the fluffy seed heads and making chains of dandelion stems. Perhaps he smiled as he watched a boy pick the stem with a satisfying snap, and examine the white pincushion top stuck full of sharp brown seeds, then blow the gray parachutes high into the air. "All children are inspired by a similar instinct," he wrote, and it's true to this day: all children know what to do when confronted with a dandelion puffball.[5] Take a deep breath and blow!

On a warm spring day, the floating seeds seem to lift mysteriously, as though under their own power; they rise on invisible columns of sun-warmed air, soaring on the thermals like migrating hawks. The tiny parachutists may travel for days, as high as an airplane, or go no farther than the neighbor's yard; sometimes a group of low-flying seeds hits a wall, and they all drop to the ground, which is why you often see a row of dandelions peeking up from the cracks where buildings meet pavement. But wherever it soars to, eventually the seed lands and starts to put down roots.

～

When Thoreau studied a subject, he did so thoroughly, and in his study of the dispersal of seeds he quite literally dug deep. "Much . . . is to be learned by examining the roots," Thoreau declared, and he proceeded to do so; he got out in the field with a ruler and a shovel and started digging.[6]

Thoreau was fascinated by the mysteries under the soil. He filled pages of his journal with painstaking measurements, recording the precise diameter and length of roots at different times of the year. He spent many days comparing plants' underground structures to the greenery that grew above them. Such research leads to a remarkable conclusion: that there is generally as much or more of a plant below the soil as there is above it. We glance at a dandelion and think that what we see is all there is; Thoreau, after delving into the matter, realized that we need to look more deeply.

Half of the biomass, or living material, of many terrestrial plants is found below the ground.[7] Almost everywhere we walk, on grass, sidewalk or paved roadway, beneath our feet there is a network of living roots, like eels swimming beneath the quiet surface of a lake. The vast tangle of roots compete fiercely, vying with each other in a slow-motion battle for space, water, and food.

In this struggle for survival, there are two basic root "game plans." Some plants go long, sending the roots to grow deep, reaching for the sure thing — the constant water supply far below. Or there's the opposite strategy, taken by other species: go shallow. The roots can spread out like the spokes of a wheel, growing quickly to form a spider web of rootlets all around the stem. This wide net can grab up the water from the merest drizzle as soon as it hits the ground, before it can evaporate.

So, which strategy does the dandelion use? The slow, patient growth of the deep taproot towards the water table, or the quick, opportunistic use of every passing shower? If you take the trouble to get out the shovel and the ruler, you'll find that it uses both. The dandelion quickly develops a wide network of shallow roots, followed by a deep taproot that is usually about a foot and a half long, but is capable of penetrating to a depth of fifteen feet (Figure 4).[8]

The general pattern of succession on barren soil is that annuals, plants with a one-year life span, move in first, produce shallow roots, and rapidly generate a whole lot of seeds. Then come biennials, plants with a two year life-span, followed by perennials, slowly but surely establishing large roots that can store up energy for the years to come. First the quick fix, then the slow healing. Dandelions are perennials, but they have many of the qualities of annuals; the versatile dandelion is both tortoise and hare.

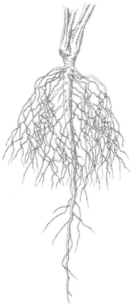

Figure 4. The root of the dandelion consists of numerous shallow roots near the soil surface and a taproot that penetrates to a greater depth.

The roots, however, are just the first step in the construction of this mechanism that works like a Swiss watch. As soon as the first rootlets have gained a foothold, leaves sprout in a tiny fountain of green. But the leaves don't reach for the sky — they lie low, creeping out over the earth. This pattern of growth, common to many lawn invaders, is called a basal rosette: the flat circle is almost immune to grazing, to being knocked over by wind, or to damage by a sudden heavy snow. You can literally walk all over the leaves, and it doesn't faze them a bit. This is an adaptation evolved eons ago, but which is uniquely suited to the modern suburban environment: as the mower blades go by an inch above them, the dandelions leaves are safe; they hug the ground, like soldiers

ducking enemy fire. Death and destruction pass by overhead;
taller plants are mowed, trampled, or beheaded, while the
dandelion leaves go placidly about leaves' eternal business
— using sun and air to make food for the plant.

To the gardener, frowning down at the dandelions scattered across the lawn, the raggedy circle of leaves may not
seem impressive. But like every part of the dandelion, the
leaves are superbly adapted to maximize the plant's chances
of survival even in dry, inhospitable places. The leaves grow
out from the center like the spokes of a wheel, overlapping as
little as possible to maximize absorption of the sun's energy.
This rosette serves as a funnel, catching rain or dew, each
leaf reaching out over the surrounding plants to steal the
precious moisture. The sides of each leaf are slightly tilted
towards the midrib, so the water slides neatly down towards
the center of the plant and is deposited right over the roots
with maximum efficiency. Inside each leaf — in fact, inside
all parts of the dandelion — is a milky, sticky sap, a sort of
Elmer's-glue-like juice that tastes bad enough to discourage
animals from eating it.

In the center of the rosette, just beneath the soil, round
green buds form. Then they wait, tightly folded, until conditions are right, and a blend of external circumstances — the
correct amount of water, daylight, and warmth, combined
with sufficient energy stored in the root — triggers the next
crucial step. The stalk shoots up swiftly, rising from the center of the rosette, and at its tip is the tightly-wrapped bud —
the possibility of the next generation.

The stem of a dandelion appears unremarkable — a
naked, leafless stalk that is a peculiar color of pinkish-gray.
But it's yet another piece of this superbly adapted mechanism. The stalk is hollow, a more structurally stable shape

than a solid cylinder of the same mass, so that the stem, fragile though it seems, is well able to withstand wind and other stress, and can fulfill its function of holding the flower head high.

The top of the stem ends in green leafy structures with a coquettish curl. These are called bracts, and they are actually modified leaves — they act as a barrier to crawling insects who might shinny up the stem and damage the delicate flower parts. Earth-bound insects would be of little use in carrying pollen from plant to plant, so the dandelion fences them out. The dandelion flower is aiming at the skies.

~

When Thoreau rambled the meadows of Concord, looking for plant specimens, he didn't have any trouble locating dandelions. The flower he described as "rich yellow"[9] has a color so intense that it almost seems to vibrate. Yellow is a vivid contrast to green; it stands out, calling in the language of color, loud and clear — not to humans, but to insects. More than a hundred species of insects are drawn to the dandelion by the scent of its sweet nectar, and by its color.

Even to human eyes, the bright yellow disk is an obvious target. Bees and some other pollinators see light and color differently than we do. In the twentieth century, scientists and nature photographers discovered that most flowers have distinct patterns that guide pollinators to their nectar — a sort of flashing neon light saying "Drink here!" — that only creatures with vision in the ultraviolet part of the light spectrum can see. We can't be sure of how these colors appear to an insect, but under UV light, dandelions' reproductive parts shine a lurid crimson.[10]

Dandelion flowers are very responsive to light. With an eerily human awareness of when the sun is out, the flower heads close up on rainy days, like an umbrella in reverse.

They also close as the sun goes down, the golden blossoms gently enwrapped by green petal-like structures called sepals that protect the delicate reproductive parts from damage — no point in displaying them while there is little chance of a pollinator fluttering past, so the dandelion closes up, like a shopkeeper shuttering the windows when there are no customers around. On clear days, the sepals open, revealing the bright yellow flowers that echo the brilliance of the sun above them. The customers flock to the merchandise, as it were. And as the bees, flies, and beetles feast on nectar, traveling from plant to plant, they inadvertently carry grains of pollen from the male parts of one dandelion flower to the female parts of another.

Now you may be thinking that a dandelion has one big yellow flower, about an inch across. But actually, the structure of the dandelion is much more complex than it looks. Each of the yellow "petals" is actually a botanically complete flower (Figure 5). Pull one out from the flower head, and you'll see that it looks like a long, thin yellow strap, which narrows into a hollow tube filled with sweet nectar. At its base is a little thread with curled tips in the shape of a Y. This is the female part, the pistil, which contains the ovary at its base. The male parts, the stamens, produce pollen, and are fused in a tube around the pistil. There are about two hundred of these flowers on a single dandelion head, arranged in concentric circles on the top of the stalk. If you look closely, you can see that some of the tiny flowers are in full bloom, with the Y sticking up, while others are still tightly folded.

So each of what we call dandelion "flowers" are really flower heads with hundreds of flowers grouped together — a highly effective strategy for success of pollination. Once the ovary of an individual flower is fertilized by a grain of pollen,

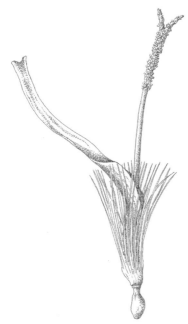

Figure 5. The flower of the dandelion. The parachute that will carry the seed is already forming at its base.

the embryonic seed begins to ripen. The sepals close up again, sheltering the babies. When the sepals open a few days later, the golden flowers have magically changed to a globe of silver.

Then another amazing spurt from the energetic dandelion: the stem elongates, pushing the launching platform higher, to reach the winds that will disperse the seeds. This is why one morning the yard contains yellow flowers pressed close to the ground, hardly an inch above the neatly trimmed grass, and before you know it the lawn is full of lanky stems topped with gray seed-heads. A whisper of breeze sends the parachutes floating high; the seeds leave the mother plant behind like the sons in the fairy-tale, and wander off into the world to seek their fortune.

Thoreau noted the short life of the dandelion flower, and seemed to feel a chill of melancholy as he noted how early in the season the yellow blossom turned to seed. "It is interesting as the first of that class of downy or fuzzy seeds so common in the fall," he wrote. "By the 4th of June they are generally gone to seed in the rank grass. You see it dotted with a thousand downy spheres, and children now make ringlets of their crispy stems."[11]

More than ten years earlier, as he left his beloved cabin on the shores of Walden Pond, Thoreau wrote that it was because "it seemed to me that I had several more lives to live, and could not spare any more time for that one."[12] He was barely thirty then, but already he seemed to feel a sense of time pursuing him. A decade later, as he was beginning *The Dispersion of Seeds,* he wrote of the ball of seeds that "boys blow to see if their mothers want them [to come home]."[13] Thoreau knew well the old New England belief that the dandelion can tell children if their mothers want them to get started on the evening chores. Perhaps he remembered when he was a boy, lying on his stomach in the sunny meadow, blowing on the dandelion puffball — you have to blow hard, since if you don't blow off every seed, it's time to head home and get to work.

The sight of a dandelion, reaching its fruition so early in the spring, made him thoughtful. "It is commonly the first of the many hints we get to be about our own tasks, which our Mother has set us, and bringing something to pass ourselves."[14] When he wrote these words about the dandelions, he was nearing the end of his too-short life; his great work on the dispersal of seeds remains forever unfinished. "We may depend upon it that our Genius wants us and always

will, till we can blow away the firmament itself at a puff,"
he added. "So much more surely and rapidly does Nature
work than man."[15] He saw the first early dandelions filling
the meadows in the spring of 1862, but he died that May of
tuberculosis, aged forty-four.

Right in the midst of his section on dandelions, a half-
page of manuscript is missing. Perhaps the lost scrap of paper will turn up in an attic or a bureau drawer someplace,
someday; otherwise we will never know what else Thoreau
had to say about the dandelion, either for praise or blame.
But we know that his study of dandelions and other common plants led him to a deeper understanding of the intricate, tangled web of relationships that exists among living
things and their environment. The writings of Thoreau pio-
neered the way to the study of these interrelationships, a
discipline that scientists now call "ecology."

Most of the Concord woods and meadows that Thoreau
rambled are pavement now. But to this day, the familiar
golden weed pops up anywhere it can in the quiet Massachusetts town; it lines roadways, speckles lawns, thrusts
through the blacktop of parking lots. Every spring, the rich
yellow is seen again, on the shores of Walden Pond, and in
cracks in the sidewalks of downtown Concord.

Look! A dandelion stalk thrusts jauntily from a crevice
in the cement, holding high a globe as fragile as smoke. It
seems to wait for one of the children of Thoreau to reach out
a grubby hand to pick the seed-head, then smile, take a deep
breath, and blow. The seeds will fly up into the springtime
sky, rising on the sun-warmed breeze, drifting off to who
knows where . . .

Notes to Chapter 3

1. Thoreau, Henry David. *Wild Fruits*. Edited by Bradley P. Dean. New York, NY: W. W. Norton and Co., 2002.

2. Thoreau, Henry David. *Faith in a Seed: The Dispersion of Seeds and Other Late Natural History Writings*. Edited by Bradley P. Dean. Washington, DC: Island Press, 1993.

3. Tilford, Gregory. *From Earth to Herbalist*. Missoula, MT: Mountain Press Publishing Co., 1998.

4. Thoreau, *Faith in a Seed*.

5. *Ibid.*

6. *Ibid.*

7. Elliott, Douglas B. *Roots: An Underground Botany and Foragers' Guide*. Old Greenwich, CT: The Chatham Press, 1976.

8. *http://www.ipm.ucdavis.edu/PMG/PESTNOTES/pn7469.html*, a website of the University of California Statewide Integrated Pest Management Program.

9. Thoreau, *Faith in a Seed*.

10. To see what a dandelion might look like to a bee, see the collection of photographs at *http://www.naturfotograf.com/UV_flowers_list.html*, a website "devoted to the visual impact of nature," created by nature photographer Bjorn Rorslett.

11. Thoreau, *Wild Fruits*.

12. Thoreau, Henry David. *Walden*. Philadelphia, PA: Running Press, 1987. First published in 1854.

13. Thoreau, *Faith in a Seed*.

14. *Ibid.*

15. *Ibid.*

Chapter 4

The Many-Headed Tribe

Lapland
1731

It's cold in Lapland, even in spring, but then a little cold weather never bothers a dandelion. A little cold weather presumably doesn't bother Karl von Linne either; the young naturalist was born in the chilly, dark country of Sweden, and wintry weather is nothing new to him.

In spite of the northern climate he comes from, he's a passionate, hot-blooded young man, at the beginning of a life-long love affair with nature, especially plants in all their fascinating and varied forms. He's trekking the hills and meadows of Lapland in search of plants for his collection, and he bends to pick a dandelion out of the frosty mud. Making sure he's got a complete specimen, he carefully gathers every root and leaf.

Later, he studies the same dandelion, now carefully dried as part of his herbarium collection. Across tables and on shelves in his study are spread dozens, hundreds, thousands of plants, of all shapes, sizes, colors. He holds the dandelion specimen between long, narrow fingers, studying the plant, pondering its secrets of form and structure. Finally he nods decisively, puts the plant down, and writes in his notebook. Karl von Linne, or to use the Latin form of his name which *he preferred, Carolus Linnaeus, has put yet another living thing into its place in the universe, bringing order to the chaos of the world, by giving it a name.*[1]

~

I once had a two-hour discussion with a fellow naturalist on whether the dainty wildflower we'd found on a woodland trail was a "small-flowered crowfoot" or a "kidney-leafed buttercup." We each consulted separate reference books, but couldn't agree; we crouched over the plant, scrutinizing the tiny leaves as we swatted mosquitoes. The discussion grew increasingly heated, with both of us triumphantly pointing to the pictures in our illustrated field guides to prove ourselves right.

But finally, we read the Latin words in fine print beneath the illustrations, and at once the mystery was solved. We were debating over a plant that had more than one common name. However, thanks to the Linnaean system of binomial nomenclature — "names using two words" — we discovered that we were looking at *Ranunculus abortivus.* Botanists around the world, no matter what language they speak, or what regional plant names are used, agree on a single Latin-based name for this little wildflower. Linnaeus had, once again, cleared things up and brought order to a confusing world.

What name do you give to, say, a long sandwich? Well, it depends on where you're from. They might call it a "sub" in New York, but a "grinder" in Connecticut; it's a "hoagie" in New Jersey, but a "torpedo" in Oregon. Kids wash it down with "soda" in Massachusetts, but they drink "pop" in Kansas. Not long ago, it was that way with plants. Almost everyone knew the names of the common wild plants that provided food or medicine — but which name?

The shrub called "shadbush" in New York is known as "serviceberry" in Tennessee. "Aaron's rod," "Quaker rouge," and "torchweed" all refer to the tall, spiky plant usually called common mullein. This isn't just a dilemma of old-fashioned flower-lovers scratching their heads over out-of-print books — to this day, if you Google *Ranunculus abortivus,"* the website of the Kentucky Native Plant Society will tell you it's a kidney-leaf buttercup, while the website of the Connecticut Botanical Society insists it's a small-flowered crow-foot — and at the Illinois State Museum site, it's the little-leaf buttercup. Go figure.

This regional mosaic of common names is especially typical of plants that have an ancient and widespread history of use by humans. Dandelions are among the plants that have the most aliases; there are more than fifty ways to say "dandelion."[2] Some are wickedly descriptive: in medieval times the dandelion was called "priest's crown" or "monk's head" because of the white and flabby bald head left after the seeds are gone. Blowball, puffball, clockflower, Irish daisy, cankerwort, and that's not even counting all the other languages of the world that have names for the dandelion. My favorite is one of the Chinese names, which translates as "yellow-flowered earth nail."

Karl von Linne was a naturalist himself, as well as a scientist, physician, and traveler in the enlightened era of eighteenth-century Europe. He grew up in Sweden, the son of an avid gardener, and even as a child he loved the green and fascinating shapes of the plants in his father's garden. But in his youth, his chosen study of botany was made baffling, and often infuriating, by the sheer number of names that could be borne by a single plant.

It was bad enough that many plants had thirty or forty common names, varying from region to region in a country or across a continent. Then each language — French, German, English, Swedish — had multiple names for the same plants. Latin and Greek were the universal languages of scientists and other well-educated persons, but all would-be botanists merrily created whatever impressive Latin or Greek name they preferred. A single species of rose, for instance, was *Rosa sylvestris alba cum rubore, folio glabro* to one botanist, but *Rosa sylvestris inodora seu canina* to the next.[3]

Linnaeus probably had his share of arguments like the one over the buttercup/crowfoot. He was a medical practitioner — he specialized in the treatment of syphilis — and saw clearly that this was a matter of more weight than merely satisfying arguments between persnickety botanists. If plants were to be used as medicine, it was obviously of some importance to know precisely which plant was being prescribed. So Linnaeus developed his system of binomial nomenclature, and spent a good deal of his long life in the tireless pursuit of naming things.

In Linnaeus's simple but elegant plan, the name of every species is made up of two words. The first word, always capitalized, is the name of the genus: *Taraxacum*, in the case of the dandelion. In the genus *Taraxacum*, there are dozens of species of dandelions.[4] The second word in the name of a species is called the specific epithet; in the case of our dandelion, this word is *officinale*. When the name of a genus and a specific epithet are combined into a Linnaean binomial, in this case *Taraxacum officinale*, they form the name of a species — a name that is unique to one particular species. There is no doubt to which species this binomial refers, whether the plant is called the Common Dandelion as in most field guides today, or swine's snout, milk gowan, or telltime as in other places and times.

The task Linnaeus set himself — to sort out and classify all life on the planet — was one that might well have daunted a lesser man. But he saw the crying need for someone to decree a clear, understandable system of nomenclature, and he elected himself to the position. With Scandinavian calm, the young man — he was only in his early twenties when he started — rolled up his sleeves and set to work.

He first divided living things into two categories: the Animal Kingdom and the Plant Kingdom. Scientists have made things a bit more complex since his day, throwing in extra kingdoms for things like fungi and the more peculiar forms of bacteria. Linnaeus, however, saw no shades of gray; with the sublime simplicity of God separating Night from Day, he simply split the living universe down the middle: Plant or Animal.

But then the work really began. The Plant Kingdom: carrots, daisies, microscopic specks of algae, towering oak trees, the baffling variations of roses and tulips, mosses, grass and ferns . . . A lesser man, contemplating the task he had set himself, might have felt as overwhelmed as Noah trying to load up the ark. An intimidating array of plants confronted Linnaeus, with a flood of new species being discovered every year as explorers investigated the New World and other distant lands. It was a problem to puzzle the wisdom of Solomon: where to begin, to bring order to such chaos?

The surface similarities of plants are deceptive. Told to sort out a group of plants, one might choose to put all the red flowers in one pile and all the yellow ones in another, or all the large blossoms in one group, separated from the smaller flowers. But Linnaeus looked deeper than the outward trappings, and ignored the distractions of color, size, or shape.

To put it in animal terms, he didn't lump, say, fish and whales together. He grouped whales — not with fish — but with mice and rabbits and all the other mammals, who bear live young. Even though fish and whales look similar on the outside, Linnaeus looked at the underlying structure, and especially at how organisms reproduce.

Perhaps not surprisingly for a specialist in venereal disease, for Linnaeus it pretty much all came down to sex. He looked through his carefully preserved plant specimens, and decided that his system of plant classification was to be based solely on the number and arrangement of reproductive organs.

I find in myself a regrettable tendency to humanize plants, and to describe their actions in human terms. I'll talk about a rose "deciding" to have thorns, or describe dandelions as wily, aggressive, or stubborn. Linnaeus, scientist though he was, seems to have fallen into the same habit; in fact, he became romantically eloquent about the whole affair of plants' reproductive lives.

> *The flowers' leaves . . . serve as bridal beds which the Creator has so gloriously arranged, adorned with such noble bed curtains, and perfumed with so many soft scents that the bridegroom with his bride might there celebrate their nuptials with so much the greater solemnity . . .*[5]

He went on to describe the intimate details of the "bridal beds," with a brutal frankness that shocked his contemporaries. He was so explicit in his descriptions of plants' male and female sexual parts that one indignant botanist, Johann Siegesbeck, referred to Linnaeus's work as "loathsome harlotry."[6]

Linnaeus proceeded with his X-rated classification scheme, unaware of the ripples of shock that were beginning to spread through Europe — or at least the part of it that was interested in botany. Accusations of indecency caught the attention of religious authorities. The Pope forbade Linnaeus's works to be housed in the Vatican libraries.

Up till this time, botany had been considered a genteel and proper study for well-bred ladies. But Linnaeus was using language that would make a sailor blush; for example, he described a plant with eight stamens and one pistil as eight men in the same bed with one bride. After this sort of thing, it's no wonder that educators began to suggest that girls and boys should not be allowed to study botany together. "Who would have thought that bluebells, lilies and onions could be up to such immorality?" huffed the outraged Siegesbeck.[7]

Linnaeus ignored this prudery, and kept at his task. He observed that there was an enormous number of plants, with very different outward appearances, that shared a common reproductive strategy. A goldenrod, a daisy, a marigold, and a dandelion may look different on the outside, but underneath the showy colors and shapes they all have the same reproductive pattern: each "flower" is actually a flower head, a composite group composed of a cluster of many small flowers, and each flower develops into a single seed called an achene. Linnaeus grouped all the plants that had this structure into a family called the Compositae. (The group is more commonly known today as the Asteraceae, or Aster Family.) He divided some families into smaller subdivisions called "tribes." The dandelion is in the tribe Lactuceae, which includes the well-known coffee-substitute chicory and a host of other plants.[8]

He had a big job on his hands when it came to sorting out the Compositae Family. It contains a huge number of plant species — there are an estimated 20,000 types of composites scattered around the planet, making it the largest plant family in the world. And perhaps — no one has actually counted — the most common representative of this family is the dandelion.

The dandelion's basic structure is replicated in every one of the thousands of species of Compositae. Look at a black-eyed Susan, or an aster, or a goldenrod. The pattern of lots of little flowers all gathered together on a single flower head is repeated again and again. On some plants, like the daisy, the flowers masquerade convincingly as petals, or hide in the fuzzy yellow disk in the center. But each of what we call daisy or goldenrod "flowers" are really flower heads with dozens or hundreds or thousands of flowers grouped together.

This is a highly effective strategy. It's arguably the most successful strategy, in terms of survival of the species, that a plant can have. Let's peer, like Linnaeus, into the "bridal beds" to examine the reproductive life of these flowers. Survival of the species all comes down to one question: how well does the species reproduce?

Plants reproduce in two ways, vegetatively and sexually. Most of the Compositae, and especially the dandelion, are adept at either one. Mammals, of course, don't reproduce vegetatively. A rat, chopped in half, is a dead rat. But a dandelion, chopped in half, is two dandelions. The new plant is an exact genetic duplicate of the original dandelion: a clone. If you want to dig out the roots of every dandelion on your lawn, you'd better be prepared for a long day's work. And you'd better dig deep. If you don't, each chopped-up bit of dandelion root that you miss could grow into a whole new dandelion. That's why you can't just pull a dandelion up by the roots and expect to be done with it; if you leave an inch-long bit in the ground, it will sprout anew. Like the Hydra who produced two heads for every one that was cut off, the dandelion root is almost impossible to kill.

Clones are headlines these days, the very cutting edge of technology, the stuff of science fiction, but plants and other organisms have been cloning themselves since early in their evolutionary history. It's a shockingly effective way to reproduce. A dandelion root chopped into ten pieces could grow into ten dandelions. Each of them can clone into ten more, and more, and so on.

So why do plants bother with sexual reproduction, then? There's a huge energy price tag attached to producing offspring, as any weary parent will tell you. Why go to all the trouble of attracting pollinators, producing seed, growing parachutes for dispersal, etc?

Well, a lot of garden plants don't bother with sex anymore. Tulips, lovingly tended by their willing human servants, often don't create viable seeds — the ability to sexually reproduce has been all but bred out of them by generations of horticulturists who want every flower they sell to be an exact duplicate of the parent. After all, horticulturists are creating a piece of merchandise that must turn out exactly the way it's represented in the garden catalog. So you don't buy tulip seed. You buy tulip clones, in the shape of bulbs that have cloned off another bulb, which cloned off another bulb, and so on back into the plant's history, all the bulbs having exactly the same genetic information.[9]

But without sexual reproduction, there can never be any new genes coming into the mix. What happens if a tulip disease comes along? If all the tulips are precisely the same, they all go down to the disease. But for a plant that reproduces sexually, with two parents providing different genetic material, there's a greater chance that one or two individuals in a population will have a combination of genes that enable them to resist that disease, or predator, or storm, or whatever adverse condition comes along. So there's an immense advantage to reproducing sexually, even if only occasionally, that makes it worth all the effort, worth the enormous price tag.

～

The pollen, which Linnaeus likened to male sperm cells, must get to the ovary of another plant, which he called the womb. Pollen must, therefore, travel. If the pollen only gets as far as the ovary of the same plant it came from, you're back to the problem of having no genetic variation. Most plants have evolved ways to avoid pollinating themselves, such as having male and female parts be of different heights (as does the dandelion) or having them ripen at different times.

But how does the pollen get from place to place? How does a plant "impregnate" another plant?

Some plants harness the wind. A single ragweed plant can create millions of tiny grains of pollen which are literally blown far and wide. (All these uncountable billions of ragweed pollen grains floating around is why ragweed is one of the primary causes of allergies. Other windborne pollens — those of the oaks, maples, various species of grasses — also contribute to hay fever.) But that's an inefficient strategy, really. There's an energy price tag to creating all those billions of pollen grains. Out of ten million specks, perhaps one, by sheer random chance, will alight on a ragweed with female parts ready to be fertilized. Perhaps there's a better way?

Perhaps, if there were a way to persuade something to carry the pollen from one flower directly to another flower of the same species, might it work better? It would have to be something that could move from place to place quickly — in other words, an animal. But why would an animal want to help a plant? There'd have to be something in it for the animal — they'd need some incentive. Like, say, food. Sweet, high-energy nectar, of great nutritional value.

But how would the animal know the nectar was available? The plant would need to put up a billboard, so to speak — do some advertising to attract pollinators to the bait. Something big, something bright. Maybe it should smell good, too.

Flowers are beautiful — at least, most humans seem to find them so. There's something about their scent, their symmetry, their color, that entices us to come closer. But plants produce flowers, not for human enchantment, but for the bewitchment of pollinators. Flowers are publicity.

Bees are common pollinators, of course, but there are also flies, bats, hummingbirds, moths, beetles, . . . thousands of species of potential pollinators — organisms that spread one plant's pollen to another accidentally, in their quest for nectar. And plants have evolved thousands of ways to attract pollinators.

Some plants use their size, their shape, or their vivid color to attract those pollinators that see well. Other flowers attract pollinators who rely on scent more than vision, so their flowers are less conspicuous to the eye, but send out a strong smell. We are most familiar with garden flowers that have a sweet, insect-attracting fragrance, but there are flowers that have a horrid stench: a graceful red trillium flower has a delightful aroma of rotting liver which very successfully attracts the flies and carrion beetles that pollinate it.

But how do the composites do it? Each individual dandelion flower is tiny, hard to see. And dandelions don't have a strong smell. How to attract a pollinator to a flower that is the size of an eyelash?

Think of a lot of people lost in the wilderness, trying to attract the attention of a passing rescue helicopter. If the people scattered, the pilot would never spot them no matter how much they waved their arms or jumped up and down. But if they were all to stand together, they would have a much better chance of attracting that pilot's notice as he glances down — particularly if they created some sort of symmetrical pattern that stands out against the random growth around them. And their chance of attracting attention is better yet, of course, if they're all wearing T-shirts of a brilliant color that is a strong contrast to their green background.

Okay, so grouping a lot of little flowers on the end of a stalk is a way to attract pollinators. But why not just have one big flower that a pollinator can't miss?

Well, a red trillium, for example, has a solitary blossom that produces one pod with a few seeds in it, which take between one and three years to germinate. But if anything happens to that lone flower, or to the pod before the seeds ripen, that's it. End of story. One shot at immortality. Perhaps it would be better not to put all the eggs in one basket? A plant has more chances to reproduce if it has more than one flower. But remember the energy price tag! A plant must photosynthesize to create energy to reproduce. One big flower is expensive, draining a plant's hard-earned resources; two are twice as much of an investment.

One tiny dandelion flower can produce one seed; two hundred flowers, crowded on a single head, can produce two hundred seeds. But the plant would have to put a huge amount of energy into a stalk that was strong enough to support two hundred big showy flowers. The more flowers the better, but if you don't want a support stem the size of a street lamp, the flowers had better be small.

So there you have it: many small, lightweight flowers, brightly colored, all clustered in a symmetrical pattern on the end of a long stalk, held high above the grass. That's the dandelion's basic pattern, and the pattern that is echoed by all the Compositae. Flowers are publicity, and the composites excel at the marketing game.

Of all plant families, none are more aggressive, more insidious, more persistent, or better at exploiting their environment than the rowdy, exuberant, cheeky bunch that Linnaeus dubbed the Compositae. They're a family in an almost human sense of the word. The many members all share some basic characteristics, but exhibit a bewildering multitude of variations on the basic theme. Think of the row of cousins at the last family reunion or wedding you attended. Tall, short, plump, lean, surly, refined, shy, argumentative, . . . they're all a bit different. But they're all part of the family, so they all have a certain basic sameness — all with shades of red hair, perhaps, or just the same look about the eyes.

Some of the dandelions' many cousins are quickly identifiable as such: the hawkweeds, or the sow-thistles, which look like little dandelions on spiky or fuzzy stems, or a big flower named goat's-beard, which has a three-inch yellow flower head and a puffball the size of an orange. But like all families, the composites have a host of members who look markedly different from each other.

There are the gentle daisies, the long-loved garden plant gone wild, and their tough stringy cousins, the black-eyed Susans. Both of these share the composite lifestyle of many small flowers clustered on a dandelion-like head, which is surrounded by a couple of dozen "petals" that aren't really petals but flowers. (They're called ray flowers, but in the black-eyed Susan, they're sterile — they produce no seed.

Their only function is to signal to pollinators.) Daisies, like dandelions, travelled the long route, seed by seed, from Asia to Europe, over to England, then across the pond to the New World. Possessing few of dandelion's medicinal qualities, their passport was their beauty — like tulips, they were passed from gardener to gardener; but unlike the passive tulip that cannot survive without the tender care of humans, daisies had a streak of dandelion toughness in them, and have always escaped the garden to edge out into the meadows and grow wild in the pastures.

In North America, the cheery white-and-yellow daisy encountered the gold-and-brown black-eyed Susan — the two species are almost mirror reflections of each other, yet they evolved on different sides of the world. The black-eyed Susan was at first strictly a prairie flower, their tough stems and prickly, almost cactus-like leaves well suited to the arid grassland. Black-eyed Susans moved eastward as the railroads began to stretch across America, and probably first bumped into their daisy counterparts in the Midwest someplace.

Linnaeus was brought specimens of black-eyed Susans by fellow botanists returning from explorations in the New World; their intense golden orange must have glowed brightly even in dried specimens, and seemed a warm ray of sunshine in the cold nights of Sweden. Linnaeus, who often named beautiful plants after his favorite people, christened the golden flower after his revered mentor, botanist Olof Rudbeck. Linnaeus wrote to his friend that "So long as the earth shall survive, and each spring shall see it covered with flowers, the *Rudbeckia* will preserve your glorious name."[10]

Tiny, daisy-like asters are called "farewell-summer" because they are the last flowers of the year to bloom. The sight of the first aster on a hot August day signals the untimely end of summer like the first "Back-to-School Sale!" sign in a mall. Frost-flower, starwort, starflower — asters have many names. There are so many varieties, hybrids, and species of asters that they must have caused Linnaeus considerable headaches, and indeed the asters cause botanists to scratch their heads to this day.

Look closely at the flower heads of goldenrod that wave like golden feathers in the autumn fields, and you'll find that each tall plume is made up of a thousand little dandelion-like flower heads. When goldenrod goes to seed, there are a thousand little gray puffballs. Goldenrods, like dandelions, excel at vegetative reproduction; a single goldenrod can spread a web of roots out 360° to create a circular mass of plants, an all-but-impenetrable thicket that can exceed thirty feet in diameter. The roots send out chemicals that inhibit the growth of other plants nearby, ruthlessly poisoning the competition. (Goldenrod pollen is carried by insects, not wind, so it does not cause the hay fever for which it is often blamed. The inconspicuous ragweed, blooming at about the same time as the showy goldenrod, is usually the culprit.)

Thistles are composites too, each purple spiky "petal" being a flower. The seeds disperse on parachutes of soft thistledown that goldfinches use to line their nests. There are many native species of thistle, but most common in pastures is the invasive, non-native Canada thistle, with small dusty-purple flower heads. It's too prickly for cows to eat, but it spreads so fast in pastures and crop fields that there used to be laws in England and Australia against allowing thistles to grow on your land lest they spread to all the neighbors' fields.

Yarrow is yet another composite. Linnaeus gave the dusty white plant the name *Achillea*, after the Greek hero of the Trojan War who traditionally used yarrow, a well-known styptic plant that slows the flow of blood, to heal the wounds of his soldiers. On the rim of each circular yarrow flower head is a row of sterile flowers that attract pollinators but produce no seed; Linnaeus unblushingly described these as "harlots," since they "mate" without producing offspring.

The list of composites goes on and on, all twenty thousand of them, wildflowers and garden flowers, from asters to zinnias. The dandelion and its cousins are, arguably, the most perfect plants in the world, if you define perfection as success in getting into the gene pool, extending the range, and perpetuating the species. Compositae are considered to be one of the most recently evolved families of plants, the very latest model.

∽

Linnaeus spent decades convincing scientists to use his system of binomial nomenclature, and he carried on a long and barbed correspondence with his adversary Siegesbeck over the propriety of using shocking words like "ovary" and "sperm" in a book that might possibly be read by females. The feud was followed by their fellow botanists with the sort of titillated interest that today might be given to episodes of *American Idol*. Irate letters, with insults couched in elegant Latin, flew back and forth. Linnaeus, always hot-blooded, was not slow in replying to criticism, frequently employing non-botanical terms such as "idiot and fool."[11]

Siegesbeck continued to protest, but Linnaeus had the last word. The Linnaean system, though somewhat modified, is in worldwide use more than two hundred years after his death, while Siegesbeck is a footnote to botanical history. Siegesbeck's name might have been totally forgotten, in fact, and it's thanks to his rival that his name will live on.

Carolus Linnaeus bestowed the name *Siegesbeckia orientalis* on a dandelion relative, a small and fuzzy creeping plant with inconspicuous flowers.[12]

He didn't give his enemy's name to the dandelion, of course. That would have been a compliment.

Notes to Chapter 4

1. Linnaeus personally collected the type specimen of the dandelion in Lapland, and that identical dandelion is still preserved in the Library of the Institut de France in Paris. Linnaeus gave it the name *Leontodon taraxacum* Linnaeus in his *Species Plantarum* (1753; 798).

2. Dutton, Joan Parry. *Plants of Colonial Williamsburg*. Williamsburg, VA: The Colonial Williamsburg Foundation, 1979. Names I have so far encountered for dandelions include Irish daisy, lion's tooth, swine's snout, door head clock, milk witch, witches' milk, blowball, milk gowan, yellow gowan, witch-gowan, priest's crown, monk's head, clockflower, telltime, fairy clock, bitterwort, peasant's cloak, wild endive, fortune-teller, dumbledor, cankerwort, and the delightfully wicked sin-in-the-grass. In French, dandelions are also known as *pissenlits*, in reference to their diuretic powers (*lit* means bed); in English, they're sometimes called pissabeds. In China, the dandelion has common names that translate loosely as flowering and hoeing-weed, golden hairpin-weed, and yellow-flowered earth nail.

3. *http://www.ucmp.berkeley.edu/history/linnaeus.html*, a website of the University of California Museum of Paleontology.

4. Sanders, Jack. *The Secrets of Wildflowers*. Guilford, CT: The Globe Pequot Press, 2003. There are dozens of *Taraxacum* species worldwide — botanists do not agree on exactly how many. The US Department of Agriculture plant database considers five species native to the US: Northern dandelion (*Taraxacum phymatocarpum*) and the intriguingly-named fleshy dandelion (*Taraxacum carneocoloratum*), both found in Alaska; the California dandelion (*Taraxacum californicum*) on the west coast; and, in the Rockies, the wool-bearing (*Taraxacum eriophorum*) and harp (*Taraxacum lyratum*) dandelions.

5. http://www.ucmp.berkeley.edu/history/linnaeus.html.

6. Blunt, Wilfrid. *The Compleat Naturalist: A Life of Linnaeus*. New York, NY: Viking Press, 1971.

7. *Ibid.*

8. In the centuries that have passed since Linnaeus named the dandelion *Leodonton taraxacum* Linnaeus — adding his own name after the plant's — botanists have debated over and fiddled with botanical names. The Compositae were renamed the Asteraceae late in the twentieth century, though the earlier name is still often used — and it's certainly easier to spell. Dandelions have had at least three other scientific names since Linnaeus. They were demeaned by the name *Taraxacum vulgare,* bestowed by Jean Baptiste de Lamarck in his work *Flore Francaise* in 1778, but most modern botanical guides ignore the *vulgare* name or list it as a synonym. The name of the familiar lawn plant known as Common Dandelion was changed to *Taraxacum officinale* by Georg Heinrich Weber in 1780 in his *Primitiae Florae Holsaticae,* and that name is generally used today. One reason for the frequent re-namings might be that the author of a revised name still has the honor of attaching his own name to that of the plant.

9. Pollan, Michael. *The Botany of Desire.* New York, NY: Random House, 2002.

10. New York Flora Association Newsletter, Vol 13, No. 1, March, 2002.

11. *http://www.scricciolo.com/linnaeus_polemic.htm.*

12. *Siegesbeckia orientalis* was used traditionally in China to treat rheumatism and is still used today in herbal cosmetics — it's supposed to help reduce stretch marks.

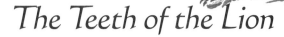

Chapter 5

The Teeth of the Lion

Shanidar, Iraq
60,000 BCE

The barren landscape is empty and seemingly lifeless;
the rocky hills are covered with a shroud of white. The sun
gleams pale and low on the horizon — it is late winter, and
although the days are growing longer, spring is still far away.
But underneath the snow, leaves are sprouting: tiny green
fingers are thrusting between the snow and the frozen earth.

A woman kneels on the ground, her short thick-set fig-
ure warmly wrapped in animal skins. A heavy shelf of brow
thrusts out over her deep-set eyes. Her strong, gnarled fin-
gers scrape away the snow, and she searches through the
browned grass to pluck the tender new growth of dandelion
greens.

A distant roar makes her jerk her head up in sudden fear.
She scans the horizon warily, then gathers up the precious
leaves and scurries off, anxious to gain the safety of her cave
before she encounters the sharp teeth of a predator.

⁓

The use of dandelions in the healing arts began so long
ago that tracing its history is like trying to catch hold of a
dandelion seed as it floats over the grass. Human memory,

never very retentive, has long forgotten who first observed that eating certain plants could make a sick or injured person feel better. But it was a simple, brilliant idea, the notion that plants had a value besides food; that roots and green leaves could sometimes affect people in ways beyond filling an empty belly, and could even make the difference between health and sickness, life and death. This idea most likely goes back a long way — long before the wheel, or agriculture, or writing.

How far back? There's no way of knowing for sure, but the unearthing of a Neanderthal grave site in a region of Iraq called Shanidar offered some tantalizing hints. In 1960, archaeologists discovered the skeleton of a man, perhaps forty years old—elderly, by Neanderthal standards. He had been buried in a cave — buried on purpose, laid deliberately to rest by his companions, and his bones had been hidden in the cave for some 60,000 years.

The skeleton was, of course, carefully studied. The meticulous archaeologists even analyzed the grains of soil around the body, and made an astonishing discovery: clusters of flower pollen. Fossilized pollen is of great interest to archaeologists because paleobotanists can sometimes identify the tiny grains, and the presence of various plant species is often used to date layers of soil.

Finding pollen in the cave was not in itself remarkable. Pollen is wafting through the air pretty much everywhere, then as now, and probably Neanderthal noses ran and eyes itched as ours do in hay fever season. But what the scientists found in this burial site were not random wind-borne specks of pollen, but good-sized heaps, in accumulations too dense to have been blown there accidentally. Therefore, they deduced, the plants containing the pollen had been placed there. And since pollen is only present in a blooming plant, this was evidence that suggested that Neanderthals had surrounded the body with flowers.[1]

A hurricane of archeological speculation followed this discovery, the Neanderthals were hailed as the first "Flower Children" (well, this was the 1960s) and the skeleton was buried anew under a blizzard of papers and research; the debate over exactly why the plants were placed around the dead man continues to this day. Of course we'll never know, but it's tempting to speculate on the funeral scene in the dark cave so long ago. Were the flowers perhaps a sign of love and sorrow for the person who had died? Were they magical offerings to appease the dead spirit? Did the Neanderthals hope for life after death? Did they have a sense of beauty?

Perhaps it was something more purposeful than beauty. Pollen of eight species of plants were found altogether, and paleobotanists succeeded in identifying six of the species present: there were hollyhocks, and a small flower like a grape hyacinth, and a type of ephedra, like a woody horsetail. And three close relatives of the dandelion: yarrow, and yellow groundsel, and a nondescript plant called St. Barnaby's thistle.[2]

Five out of the six identified species have a strong tradition of medicinal use and indeed are popular in herbal medicine to this day.[3] Some of them, like ephedra and St. Barnaby's thistle, are not particularly decorative; they are neither showy nor colorful. This information raises the intriguing question: did the Neanderthals who placed the plants in the grave understand their medical benefits? Were they the first physicians to use herbal remedies?

All this can never go beyond speculation, of course. We'll never know who first nibbled the bitter, jagged leaves of dandelions and took note of their medicinal properties. It was many, many thousands of years before someone would think to write the information down. It's more than likely that the ancient Greeks and Romans, including Pliny, used dandelions medically, but they used non-descriptive names that can't be conclusively proved to refer to dandelions. The first written reference to the familiar plant is in a medieval Arabic text of around 1100 CE.[4] By then it was a long-established medical plant, an official remedy, as its multitude of names tell us.

The word *Taraxacum* has ancient roots, though its origins are lost. It is possibly from the Greek *tarassein,* meaning "to alter," referring to the plant's effects on the body, and its ability to alter the symptoms of disease. Or it may be a Latinized version of the Persian words *talkh chakok,* meaning "bitter herb." The word *officinale* is from medieval Latin and was a name often given to plants that were of use to man; it refers to being "of the office" or "of the apothecary."[5]

Then there's the common name by which most of us know the little plant: dandelion, a pleasing name, sounding trippingly on the tongue. It also comes from medieval Latin: *dens leonis,* meaning "tooth of the lion," and dandelion was a common plant in the medicinal gardens of medieval monasteries. In French, "lion's tooth" is *dent-de-lion,* and when the Normans invaded England in 1066, they brought many French words which became Anglicized; *dent-de-lion* became "dandelion." The name of this common plant refers to lions in a similar way in almost every European language: *diente de leon* in Spanish, *Lowenzahn* in German, and so forth.[6]

But now here's a perplexing question: Why is this plant connected with, of all things, the lion? When I look at the lowly dandelions sprawled over the lawn, they might make me think of goldfinches or yellow butterflies, or some other pretty creature, but lions are not the first thing that leaps to my mind. Surely this is a humble plant, quite literally trodden underfoot?

Every reference book I consulted seemed a little unsure of itself, uncertain about exactly why these small yellow flowers are named after lions, and every herbal expert contradicts the last. Some say the words "lion's teeth" refers to the leaves, which are often sharply jagged — though dandelion leaves are quite variable, depending on where they grow: like snowflakes, no two are exactly alike, and not all of them are sharply toothed.[7] Other herbal references say it's the roots that are tooth-shaped and white when peeled. Or maybe the head of shaggy gold flowers made someone think of a lion's mane.

But lots of plants have jagged leaves, pointed roots, or yellow flowers. Why not dog's tooth or cat's tooth or something a bit more modest? Lions have a reputation to uphold, an ancient symbolism; lions have always stood for power. Why is this lowly plant exalted with the name of the King of Beasts?

To explore this question, we have to go back to the beginning — way back — to the very beginning of the struggle for survival on this planet. For the Neanderthals, as for all animals, winter was a time of death. Survival was a constant struggle, and the hardest time of all was late winter. On the calendar, the vernal equinox is in mid-March, and it's supposed to be spring then, but those of us who live in the northern latitudes know that winter often keeps its icy grip well into April. Though the sun is a little stronger and the air may be a tad warmer, the food supplies that were available in fall are exhausted; game is thinner, warier, and harder to catch, and the seeds, nuts, and berries are long gone. Nature's larder is bare.

Even after humans developed the art of agriculture, late winter was still the tough time. Through ancient millennia up to modern times, it was the same story — the months we now call February, March, and April were the hardest time of all, cold and barren in spite of increased daylight. "When the days begin to lengthen, then the cold begins to strengthen," says the farmers' proverb. Farm animals were thin, hens stopped laying eggs, the cow went dry. The apples were all eaten, the beans and squash were gone, the potatoes were rotten.

This was the time when disease would begin to make inroads in communities. Even today, early spring is the cold and flu season, with stuffy-nosed patients hacking and sneezing in doctors' offices. Not too long ago, funeral parlors and undertakers did a brisk business in the cruel end of winter.

Late March, a raw wind blowing. The pond is still frozen solid; the trees are bare. But even in the "dead" of winter, the earth is full of life. Beneath the black lid of pond ice swim turtles and fish; inside the shriveled buds on each bare branch are infinitesimal leaves, next summer's greenery already formed in miniature. Even under the snow, many plants are busily working, capturing any stray gleam of sunshine to use in the food-making process called photosynthesis.

Sunlight, even a pale winter ray, is the start to this process, the key that turns the engine, so to speak. Only a small trickle of light can penetrate the snow, like the gleam of a candle in a dark room. But even through a layer of snow that is six inches deep, enough sun can penetrate to trigger the light-sensitive cells in a dandelion leaf and start the process. Sunlight powers the long chain of chemical reactions that occurs in every green leaf, turning air and water into sugars that nourish the plant — and anything that eats it.

"A blanket of snow" is an often-used metaphor, and it's an accurate description. A heavy layer of snow insulates the soil, keeping it warmer than the air. Dandelions are one of the very first plants in the Northern Hemisphere to grow under the white covering, their leaves creeping flat on the ground between the snow and the soil. Animals sense this; deer, rabbits, moose — and in Neanderthal times, woolly mammoths — swept away the snow to find the life-giving green underneath. Despite the risk of a few leaves being nibbled, this early start is a tremendous benefit to the plant, giving it a head start on the competition from other plants — the loss of a few leaves is a small price to pay for getting a jump on spring.

Winter seems to be the longest season, lingering on long past its proper exit date. The cold and the darkness wear you down, the snow turns to dirt-crusted heaps but refuses to melt entirely, and there's mud everywhere. People always look longingly for signs of spring: the first returning goose, the first butterfly, pussy willows budding, and the maple sap-run have long been hailed with joy. An ancient spring tradition in many European cultures, which settlers spread to the New World, is to go out early in spring and pick dandelion greens. For example, there's a Pennsylvania Dutch tradition that if you eat a dandelion on Maundy Thursday (the Thursday before Easter) you'll be healthy all that year, and restaurants in that area serve spring-time dandelion greens to this day. The leaves were picked, not only as ingredients for a tasty salad, but as a spring tonic, a cure for the diseases that were sapping the strength, a way to cleanse the blood of winter's ills. It was a sort of botanical chicken soup, prescribed by anxious mothers for a long list of ailments. Why, people believed that dandelions would cure almost anything:

baldness, loose teeth, stiff joints, sores, fevers, you name it. And — believe it or not — they were right.

For many centuries, humans were puzzled by the causes of most sickness. What made a healthy person sick? Was the illness caused by demons, or the next-door-neighbor casting a spell? Was it an excess of bad humors in the blood, or the harmful effects of night air? Was your sickness a punishment for something you'd done, or the whim of the Fates, or perhaps just the inscrutable will of God?

The word "scurvy" doesn't come much under our notice today; it's a rather quaint, archaic word, like "galleon" or "pantaloons." But in its time, scurvy was as dreaded a word as AIDS is today. It was a baffling disease, which often plagued communities that were isolated for long periods of time, with little or no fresh food — crews of ships on long voyages, or snowed-in polar explorers. Or people — like the Pilgrims — who were subsisting through the winter by eating dried, salted, and pickled food. Since the early symptoms of scurvy are mild, the disease often went undiagnosed for long periods of time. Some scientists argue that scurvy has caused more death, suffering, and shortening of the human life-span than any other disease, and it continues to plague the human race to this day.[8]

Well into the 1900s, scurvy was one of the diseases, like rickets or pellagra, whose causes were unknown, or only beginning to be guessed by a few scoffed-at scientists. These diseases struck inexplicably, with no apparent pattern, except that they often tended to strike the poor. The symptoms didn't seem to be contagious, but they were lethal. Vitamin deficiency diseases have a variety of slow and subtly unpleasant symptoms, such as loosening teeth and hair loss. In their later stages come deadlier effects: depression, skin sores, and joint pain; finally, death. Unless treated, scurvy was, and is, always fatal.

Vitamin deficiency diseases could also combine with other ailments to give a deadly one-two punch. When the body is at its lowest ebb, weakened by poor nutrition, even a cold or a touch of flu can kill. The most common cause of death for the Pilgrims during their first terrible winter in New England was what they called "General Sickness," which was most likely scurvy, combined with pneumonia and tuberculosis; in the first four months in the New World, almost half of them died of it, most of them in February and March.[9]

As early as the 1600s, some scientists had observed that fresh foods were the cure for scurvy. In 1747, a Scottish naval surgeon named James Lind had discovered that consuming citrus fruits kept the symptoms of scurvy at bay. The Royal Navy experimented on its scurvy-ridden sailors for a century and a half, trying lemons, lime juice, and sauerkraut, but well past the start of the twentieth century, some doctors were still blaming scurvy on pet theories: lack of exercise, tainted canned goods, or low morale. No one seems to have understood the underlying cause of the disease — that it was not the presence of some unknown germ or bacteria that caused scurvy, but the *absence* of something.[10]

It wasn't till 1905 that a scientist, William Fletcher, discovered that if special factors were removed from food, disease occurred. Finally the scientific community caught on to the idea that there were invisible, untastable things that were removed by certain ways of processing food, such as drying, canning, salting, or over-cooking. Without these components of food — which we call vitamins — health was impossible.

Here lies the secret of the lion's power. Vitamin C is only found in fresh food: green leaves, fresh fruits, raw or under-cooked meat. In winter, most people — especially the poor — subsisted on pickled, dried or salted foods. Vitamin C cannot be synthesized by the human body, or stored in the body for long periods of time. In the twenty-first century, we're used to seeing mangoes and papayas in stores, brought to us from the ends of the earth; strawberries and watermelon are on the grocery store shelves in January. But in recent memory — in my grandmother's day — an orange was an eye-popping treat that children hoped Santa might leave in their stockings. The word "vitamin" didn't exist until the early 1900s; Vitamin C wasn't artificially synthesized, so that it could be sold in pill form, till 1935.[11]

Dandelions are a vitamin powerhouse: 100 grams of raw dandelion greens have 14,000 international units of Vitamin A, plus 35 milligrams of ascorbic acid, or Vitamin C.[12] That's more Vitamin C than tomatoes, and seven times the Vitamin A of oranges, pound for pound. In addition, dandelions have significant amounts of protein, iron, thiamine, riboflavin, calcium, and potassium, as well as vitamins D, K, and B-complex. Nutrient-wise, dandelions are as powerful as . . . as lions.

As soon as a patient has fresh food, the symptoms of scurvy generally vanish with amazing speed. A blast of Vitamin C, combined with the dandelion's arsenal of other nutrients, would soon set scurvy to flight, and a mess of dandelion greens or a cup of dandelion tea would help to beat back a host of vitamin-deficiency symptoms: loose teeth would tighten in the gums, hair would grow back, joints would stop aching. More subtly, depression would lessen, and morale would improve. Dandelion remedies would greatly enhance the patient's ability to fight off many kinds of diseases.

Other plants have nutrients, certainly — strawberries, for instance, have far more Vitamin C than dandelions. But dandelions not only contain nutrients, they offer them at the crucial time — in the early spring, when the body is starved for them. Before the days of grocery stores stocked with year-round cartons of orange juice, there weren't a lot of other nutritional possibilities out there in late March. By the time the June strawberries ripened, it might be too late.

The ancient wisdom of eating dandelions in early spring has saved countless lives, and could have saved more; throughout history, many, many people must have died of scurvy and other vitamin-related diseases, and been buried in early spring, under the dandelions. The nutrients that might have preserved their lives were within easy reach — if only they had known that there was medicine under the snow, green in the very nick of time.

Notes to Chapter 5

1. Solecki, Ralph S. *Shanidar: The First Flower People*. New York, NY: Alfred A. Knopf, 1971.

2. *Ibid.*

3. Overview: History of Herbal Medicine available at: *www.naturalhealthvillage.com/reports/rpt2/oam/herb.htm.*

4. Grieve, Maud. *A Modern Herbal*. New York, NY: Dover Publications, Inc., 1971. First published in 1931.

5. Dutton, Joan Parry. *Plants of Colonial Williamsburg.*

6. Grieve, *A Modern Herbal.*

7. In John Gerard's classic *Herball*, first published in 1597, and amended and reprinted in 1633, he describes the wide variations in the jagged leaves of dandelions. In a passage unintentionally rendered hilarious by the strange seventeenth-century printer's custom of often substituting "f" for "s," he explains that dandelion leaves frequently vary in "largeneffe, fmallneffe, deepeneffe, or fhallowneffe of the divifions of the leafe, as alfo in the fmoothneffe and roughneffe thereof."

8. Stone, Irwin. *The Healing Factor: Vitamin C against Disease.* New York, NY: Grosset and Dunlap, 1972.

9. Willison, *Saints and Strangers.*

10. Huntford, Roland. *The Last Place on Earth.* New York, NY: Random House, 1983.

The famous British naval officer, Captain James Cook, was impressed with James Lind's observations, and Cook succeeded in eradicating scurvy on his ships' long voyages. But in later years, the Royal Navy bureaucracy lost faith in the power of expensive fresh citrus; they switched to bottled juice as a cost-cutting measure, thereby losing much of the fruit's effectiveness. As late as 1911, scurvy was still being blamed by some physicians on tainted or improperly canned meat.

11. The word "vitamine" was coined by Polish scientist Casimir Funk, in 1912. "Vita" for life and "amine" from the thiamine he had isolated from rice husks. Together Funk and an English biochemist, Sir Frederick Hopkins, formulated the vitamin hypothesis of deficiency disease.

12. Gibbons, Euell. *Stalking the Healthful Herbs.* New York, NY: David McKay, Co., Inc., 1971. Euell Gibbons, wild foods guru of the 1960s, wrote this little-known sequel to his famous hit *Stalking the Wild Asparagus.* In *Stalking the Healthful Herbs* he presented a good deal of nutrition information. He arranged for fourteen species of wild plants to be evaluated for their content of protein, Vitamin C and carotene, and discovered that 100 grams of dandelion buds have 3.1 grams of protein, in addition to 30 mgs of Vitamin C and 800 I.U. of carotene, also known as pro-Vitamin A. Gibbons also quoted from *The Composition of Foods, Agricultural Handbook No. 8* published by the Agricultural Research Services of the US Department of Agriculture.

Chapter 6

Swine's Snout
and Dead Men's Fingers:
A Meadow of Magic

Amsterdam, New York
2006

A little boy scampers across the field, barefoot in the warm grass. He flings himself on his stomach to roll down a hill, and lands sprawling at the bottom of the slope, arms spread wide. Lying on his back, he idly picks a dandelion puffball and blows on it, and the seeds fly high into the summer blue.

I brush the sandwich crumbs from my lap and get up from the blanket spread over the sunny grass. Smiling, I stroll over to the small figure. "Time to go pretty soon," I say. "What are you doing, honey?"

He rolls over and blinks up at me warily, his eyes alight with mischief. "Oh, nothing," he says.

I pick up a dandelion and blow on it, too, just to let him know it's all right. "You know what that means?" I inquire, as we watch the gossamer parachutes drift like smoke.

He scrambles up and brushes the grass off his Pokemon T-shirt, leaving green stains. "Sure, Mom," he says, humoring me.

"Well, what?" I persist, curious. "What does it mean if you blow on a dandelion?"

"It means the seeds are dispersing," he answers impatiently. He's the child of a naturalist, all right. I look at him, a little taken aback.

"Well, yes, that's true," I admit. "But a dandelion can tell you what time it is. Didn't you know that?"

"Huh?" he murmurs, idly hopping up and down.

"Or it can tell you who you'll marry, or how many children you'll have, or give you wishes . . ."

"Gotta go, Mom," he says, and is off, like a dandelion seed on the breeze.

∽

When I was little, someone told me that blowing on dandelion puffballs can tell you if it's time to go home and do the chores.[1] Or how many years it will be till you get married — or if your boyfriend is thinking about you . . . and of course the number of seeds left after three puffs reveals how many children you'll have.[2] Seems like everyone knew those things, when I was a kid. All the other kids did, anyway. My neighborhood gang knew love charms and superstitions and good luck techniques for a dozen different kinds of plants, way back then. But the times they are a-changing.

How about you — do you know anything about the ancient lore and magical properties of plants? No? Well, what plant can tell you if he loves you, or loves you not? Daisies, right? I'm certain that you know the correct procedure: pick the petals and drop them one by one, chanting "He/she loves me, loves me not, loves me, . . ."

What does it mean if you find a four-leafed clover? You know, don't you? For many years now, I have been conducting an informal survey on this important topic, at picnics and playgrounds and other outdoor venues. Everyone I ask stares at me like I'm asking them what planet they live on, and immediately give the correct response. I have, so far, never met anyone, man, woman, or child, of any nationality, who did not know that finding a four-leafed clover brought good luck. Almost no one has actually *found* one, mind you, but they do know what it means.

And what do you do, when you want to make sure that you can fend off bad luck? When you recklessly make a statement like "I'm gonna ace that test, for sure!" and then realize that you're tempting fate? Do you follow the ancient Druidic custom of appealing to the powerful magical spirits found in trees? Sure you do. You knock on wood.

Now ask yourself, where did you learn all that stuff about clovers, daisies, and knocking on wood? Did you read it in a book? See it on TV? Were you taught it at school, or did you get it off the Internet? Probably not. Probably, someone told you.

Someone told you. A mother, a grandmother, or another child, back when you were a kid. And where did *they* learn it? Someone had told them. That's how most knowledge has been transmitted through most of human history. Four-leafed clovers and daisies seem to be the very last remnants of a venerable, once-rich tradition, a wealth of knowledge and legend and myths about plants, that has been passed on verbally, over countless generations.

Like the use of plants in medicine, it's impossible to pin a date on these legends — they go back, to use the old phrase, "time out of mind." Daisies, for instance, were considered sacred to the Greek goddess Aphrodite, and have been used in love charms and potions for millennia. The origin of "lucky clovers" has been traced back to the Druids, a priestly cult of ancient Britain, who saw the three-in-one leaf as representative of Earth, Sea and Sky, so that clover was regarded as a powerful, magical plant. Later, Christianity put its own spin on the firmly-rooted pagan traditions, so that when St. Patrick explained the concept of the holy Trinity to the Irish, he is said to have used the clover leaf as a living illustration of how something could be three, yet one. A fourth leaf provided extra good luck, perhaps because it had the shape of the cross.[3]

From ancient times, a wealth of plant lore, healing wisdom mingled with superstition, was passed down largely by word of mouth. Today, as soon as we hear an interesting fact or important message, it's second nature to write it down. We write everything, from business e-mails to a scribbled grocery list: "Buy cat food and toothpaste," and so on, and our crowded memories have lost a good deal of their ability to remember things. Gone are the days when traveling bards could memorize all the tens of thousands of words in the *Iliad* and the *Odyssey*, and recite any section on demand, like a human CD.

In past centuries, paper wasn't an item the average person was likely to have handy. Paper was an expensive luxury, and writing was a skill only an elite few had mastered. Whether you were a silversmith's apprentice or a novice charioteer, you probably didn't learn your trade from a textbook. The skills of life were taught out loud, face to face. Someone told you.

This was particularly true of the household skills of women, few of whom could read or write until modern times.

Most cookbooks and household manuals didn't come along until the 1800s. Plant lore, especially as it related to the healing arts, was generally taught verbally. And since the responsibility for the family's health was the woman's, it was often passed down from mother to daughter or granddaughter, from aunt to niece.[4]

In towns and villages across Europe — and later, in America — there were generally a few women in the community who were known to be especially skilled in using the plants they picked from the ditches and fields as cures. Sometimes people scoffed at the "old wives' tales," but the woman who used yarrow leaves and spider webs to stop bleeding, put moldy bread on an infected wound, or mumbled charms over a cup of dandelion tea sometimes got surprisingly successful results — perhaps more successful than those of the learned physician who opened a vein in his patient's arm to remove a quart or so of "tainted blood." In the post-penicillin era, the use of mold to prevent infection doesn't seem quite the silly superstition it did to doctors in earlier times, and cobwebs and yarrow really do have styptic properties that slow the flow of blood.

But every healer loses patients, and sometimes, things didn't work out. If the patient died, or the neighbor's cow stopped giving milk, or the children got sick, whispers and rumors would begin to fly like bats. "The People . . . gave pretty free Scope to their imaginative Powers," wrote Samuel Drake, an early American chronicler, "And what one fancied or dreamed, and told to his Neighbour with an ominous Shake of the Head, was by that Neighbour told to another under a full Belief that it was true."[5] Soon the herb-wife might find herself standing in the dock, chained hand and foot, accused of the most dreaded crime of ancient times.

~

Eye of newt and toe of frog,
Wool of bat and tongue of dog,
Adder's fork and blind-worm's sting,
Lizard's leg and owlet's wing . . .
Double, double toil and trouble;
Fire burn and cauldron bubble.[6]

Dandelions have long had their place in the arts of witchcraft. Witches are infamous for using wolf's paws and cat's eyeballs and so forth in their potions, but some of these macabre ingredients are merely plants in disguise. The witch who used, say, dead men's fingers and swine's snout in a brew may have been a skilled herbal healer: dead-man's-fingers is the common name given to a type of fungus (*Xylaria polymorpha*), and swine's-snout is another of the many aliases of dandelions, from the snub-nosed shape of the closed seed head.[7] A few charms and spells muttered over the cauldron, or some spooky extraneous ingredients tossed in, could boost the patient's morale and increase confidence in the medication — which might indeed help the healing process. And undoubtedly there were herbalists who believed that they were really practicing magic.

Dandelion's milky sap was seen as representing both semen and mother's milk, so it was an important part of fertility charms. It was frequently known under other names, like witches' milk or witch-gowan, *gowan* being an archaic word for daisy. And, while planting a triangular pattern of dandelion seeds in a magical pot won't do much for your fertility, ingesting a plant rich in vitamins and protein might well provide the added nutrition that could increase a woman's chances of conception, or of giving birth to a healthy child.

Scratch a myth and find a fact, as they say. Take St. Johnswort, for example, a common plant of European meadows and pastures that made an unrecorded jump across the water to America centuries ago. The plant has bright yellow petals and an unusual reddish sap that will ooze out of a crushed leaf and stain your fingers as though with blood. The blossoms open about the end of June in a sunburst of yellow petals, just at the time of the summer solstice, Midsummer Eve. In some areas of northern Europe, the red sap gave the plant the name Balder's-blood, Balder being a Norse god who died tragically. The plant has long been considered a symbol of the sun, and has an ancient history of being a powerful protection against evil.[8]

As Christianity spread through Europe, many of the ancient pagan traditions didn't die, they merely changed their outer garments and hid among the converts to the new religion. The sunny yellow plant with the red sap became St. John His Wort, or St. Johnswort, *wort* being an old English word meaning plant. The ancient rites and celebrations of the summer solstice were metamorphosed into St. John's Eve, and the red sap became a reminder of the blood of the martyred saint.

St. Johnswort was — and is, in the Wiccan tradition, to this day — highly recommended for use in spells to counter black magic and fight off evil spirits. You can buy St. Johnswort, along with dandelion and other magical herbs, as well as cauldrons, witches' brooms and dragon's egg candles, at any one of innumerable magical supply shops or websites.[9] And, of course, if you walk into any pharmacy today, you can also buy St. John's Wort as a drug; doctors prescribe it for depression.

It's not hard to see how a plant that is medically classed as a mood elevator came to be labeled as one that fights off

evil spells. You feel sad, depressed, under the weather —
possibly evil magic is dogging your footsteps, or someone
has put a curse on you. You visit the local witch-wife, and
she brews you a potion of St. Johnswort tea. You feel better!
The bad luck or black magic that has been affecting you is
vanquished by the sun-power of the plant. Some of these ancient customs flourished for so long because there was a strong
bedrock of science under the layers of superstition and ritual.

Witchcraft and herbalism twined around each other for
centuries, like the intertwined snakes in the ancient medical
insignia of the caduceus. Learned physicians used herbs, too,
but cloaked the identities of their ingredients in Greek and
Latin names. Witchcraft was a deadly serious crime. Science and religion for once combined forces against a common target, and both the medical community and the Church
shared a vindictive hatred of witches. Over the centuries,
tens of thousands of herbal healers and midwives were
hanged, burned at the stake, or subjected to tortures of stom-
ach-churning horror.[10] Small wonder if fewer and fewer people
dared to fool around with love potions or good-luck charms,
or herbal medicines.

The knowledge of plant lore, burdened with the stigma
of witchcraft, began to wane. As the Enlightenment of the
eighteenth century brought the witch-burnings to an end, it
also brought more effective drugs and medical techniques that
pushed plants' contributions further into the background.
Medicine became a science accessible only to the highly edu-
cated, not something for illiterate old crones with a cauldron
full of magic plants to meddle with.

But as the knowledge of herbal lore faded over the centuries, a few shreds of the magic were remembered, even
though they were laughed at as superstition, never taught

formally, and rarely even written down. These traditions were kept alive, whispered in corners, passed from person to person, by certain people who aren't all that interested in what schools have to teach, and who often engage in the practice of keeping secrets from authority figures. Plant lore became the domain of children.

Kids — of course! Who else would listen with open minds and hearts to tales of magic plants, evil spells, and lucky charms? All the terror of black magic, the power of love potions, the life-and-death knowledge of healing plants, passed into children's tales and rituals. Children listened wide-eyed to the fairy-tale of the prince who sought the magic blossom at the water's edge to be cured of his wound; the witch who grew the plant in her garden that so bewitched Rapunzel's mother that she sold her child for it; the leprechaun stories of fern seed that made you invisible, and pots of gold buried under the roots of enchanted flowers. It was no longer the adults, but the children, who spent their time — and still do — looking for magic in the grass. Searching for four-leafed clovers. Counting daisy petals. Knocking on wood.

The tide of modern life has swept most of our knowledge of plants away. The plants that had the most superstitions attached to them tended to be the ones that were regarded as the most powerful and effective, and they are the ones, like daisies and dandelions, that still have a few lingering traditions, secrets passed from mouth to mouth, in an unbroken chain over the dusty centuries. I like to think of a pagan Briton mother, in woolen cloak and sandals, telling a little boy that a dandelion puffball will predict how many children he'll have. And he tells his friend, and his little brother, and then, years later, he relays the tale to his granddaughter, and so on and on, down the ages, to me. I remember myself as a child, long braids hanging down as I lay on the summer grass, blowing on a dandelion and carefully counting the seeds.

To this day, the dandelion seems to be the flower earmarked for children. In modern times, the dandelion has been adopted by an organization called the Military Brats Registry, dedicated to the children of military personnel, as their official flower. Their motto is "Children of the world, blown to all corners of the world, we bloom anywhere."[11] Also, there is a Crayola crayon of a deep, rich yellow hue that the famous crayon company calls "dandelion."

In a neatly-mowed park or a well-tended garden, the dandelion is the only flower a kid can pick and not get into trouble. Perhaps that's another reason there are so many traditions about dandelions that still linger, when the knowledge of most plants has died away.

Why, you can do no end of things with dandelions. Weave a golden crown of flowers. Pop off the heads and make a circle of the stem by putting the narrow top into the wider hollow of the bottom, to make a long, sticky necklace. Blow off the seeds, and they will float in the direction of your lover. Take a deep breath and blow again to find out if he's thinking of you — if there are any seeds left, then he is. A child in a field full of dandelions need never run out of things to do.

And of course, if you catch a flying seed, you can make a wish. But you have to let it go, for the wish to come true.

～

I stand watching Timothy dart up the hill like a dragonfly; he rolls down the slope, then climbs back up to roll down again. I wave to him, beckoning.

"Aw, Mom," he says. "Time to go home already?"

"Let's see," I answer, and bend down to pick a dandelion. "Come on over here, I've got something important to tell you."

Notes to Chapter 6

1. Botkin, B. A., ed. *A Treasury of New England Folklore*. New York, NY: Bonanza Books, 1947.

2. Silverman, Maida, *A City Herbal*. New York, NY: Alfred A. Knopf, 1977.

3. Sanders, *The Secrets of Wildflowers*.

4. Karlson, Carol. *The Devil in the Shape of a Woman; Witchcraft in Colonial New England*. New York, NY: Penguin Books, 1987.

5. Drake, Samuel. *Annals of Witchcraft in New England*. New York, NY: Benjamin Blom, 1967. First published in 1869.

6. Shakespeare, William. *Macbeth*. Edited by Robert S. Miola. New York, NY: W. W. Norton and Co., 2004. Act IV, Scene 1, lines 14-17, 20-21.

7. Buckland, Raymond. *Witchcraft from the Inside*. Saint Paul, MN: Llewellyn Publications, 1971. The ancient name of swine's snout also has Biblical ramifications: "As a ring of gold in a swine's snout, so is a fair woman without discretion." (Proverbs 11:22).

8. Wheelwright, Edith Grey. *Medicinal Plants and Their History*. New York, NY: Dover Publications, Inc., 1974. First published as *The Physick Garden: Medicinal Plants and Their History* in 1935.

9. See *http://www.goddesscircle.com/magickit.html*, and many other magical supply websites.

10. Briggs, Robin. *Witches and Neighbors*. New York, NY: Penguin Books, 1996. Recent estimates place the number of executions for witchcraft in Europe between 40,000 and 50,000. Women made up about 75% of the total.

11. See their website at *http://www.militarybrat.com*.

Chapter 7

Dr. Diligence and Mr. Honesty

London, England
1652

An Englishman named Nicholas Culpeper is sitting at a desk piled with papers. He picks up his quill pen, dips it in the inkwell, and begins to write diligently; the only sound in the candle-lit room is the scratching of his pen. He is a young man, but the warm light flickers on a face that is worn and lined, with deep shadows under the large dark eyes — he was badly wounded, not long ago, in the bitter civil war between the defenders of the English Monarchy and the Puritan reformers, and he has known what it is to suffer ill-health.

He writes intently, painstakingly recording all the knowledge he has absorbed in his short life, all his experience of the arts of healing. His skill has saved many lives, eased much suffering. He is eager to share what he has learned.

He starts off his work in bold capitals: TO THE READDR. He dips the pen in the inkwell again, gives it a shake, and continues:

Courteous Reader,

Aristotle, in his Metaphysicks writing of the Nature of Man, hit the Nail on the Head when he said, That Man is naturally enclined to, and desirous of Knowledg . . . The Subject which I here fixed my thoughts upon is . . . the Description and Nature of Herbs, which So precious hath the knowledg of the Vertues of Herbs been in former times . . ."[1]

His pen is busily scratching at the paper, but his work is interrupted by a banging at the door. Sighing, he puts down his pen and goes to answer the summons. Another patient — he has seen more than thirty already today, but he cannot refuse to treat one more.

❧

I first made the acquaintance of Dr. Culpeper, not in the dusty halls of a quiet library or bookstore, but on the Internet. He seemed rather out of place there, and his antique words and quaint phrases, written more than three hundred years ago, looked odd on a computer screen. But the entire text of his classic *The English Physician* is available on-line. Good advice never goes out of date.

His writing style is a trifle old-fashioned, but in his day young Nicholas Culpeper was a radical, a fiery rebel. As an apprentice, he started out with the usual prescribed course of study for physicians, the traditional Greek and Latin, but he soon set out to blaze a trail — or rather, to re-trace an old, old road: the country lanes and by-ways where the dandelions and other humble plants grew.

His work rocked the medical establishment. He infuriated the College of Physicians when he had the temerity to translate the *London Pharmacopoeia*, their official listing of medicines, into the vernacular. He wrote all his works in

English, so that any Tom, Dick, or Harry — or even, perchance, a female — who had enough education to spell out his forthright words could learn the healing arts.

Culpeper turned his back on the medical establishment and its emphasis on invasive, aggressive treatments like phlebotomy (the art of drawing blood from patients) and their use of chemicals and harsh drugs. He condemned in no uncertain terms the use of rare and expensive items that only the wealthy could afford, and looked in the fields and ditches for the common plants that grew where anyone, rich or poor, could pick them. He explained in layman's terms how to use herbal remedies, writing down the lore that had been passed along verbally for countless generations, but which in his day was already in danger of being forgotten.[2]

Culpeper began his career conventionally enough, attending, in his youth, the venerable Cambridge University to study Latin and Greek, and his family expected that he would enter the Church. But, as young men often do, he abandoned the textbooks when he fell in love. Being enamored of a "beautiful lady" whose disapproving father was "one of the noblest and wealthiest in Sussex," Culpeper planned to elope with her.[3] His beloved was impatiently waiting for him at the appointed trysting place (with a collection of her "rich jewels") when a storm came up unexpectedly, and she took shelter under a tree; the unfortunate lady was struck by lightning, "exchanging her marriage on earth for one in heaven."[4]

The devastated Culpeper, seeking in his grief a reason for the cruel, random accident, attributed the disaster to the "malevolence of Mars and some other envious planets."[5] He abruptly refused to return to Cambridge, and announced that he was going to study astrology and become an apothecary: two professions that often went hand in hand in the days when science blended with superstition. His subsequent books made much use of astrology. Seeking, like Linnaeus and many others before and since, to find some order in a seemingly chaotic universe, he decided that "Sickness and Health were caused Naturally (though God may have other ends best known to himself) by the various operations of the Macrocosm . . . therefore he that would know the Reason of the operation of Herbs must look up as high as the stars: I alwaies found the Disease vary according to the various motion of the stars . . ."[6] Dandelions had long been considered to be under the influence of Jupiter, largest of the planets and named after the Roman king of the gods.

In spite of the astrological trimmings, Culpeper's book *The English Physician* became a classic, widely used for more than a century after his death. His clinical experience with patients — he often treated dozens in a day — combined with his acute observation skills, gave him a good sense of what herbs worked for what diseases, even if he thought the planets and the stars were the causes of their efficiency. He explained himself and his purposes right up front in the majestic title, which is *The English Physician: or an astrological Discourse of the Vulgar Herbs of this notion; Being a Compleat Method of Physick, whereby a man may preserve his Body in Health; or cure himself being sick for three pence charge, with such things only as grow in England, they being most fit for English Bodies.*

∽

The English Physician is very user-friendly. Unlike previous herbals, which listed hundreds of plants and were the approximate size and weight of an anvil, Culpeper's book was a light-weight volume, affordable and easy to read, concentrating on those plants that might grow in the backyard.

It contains much helpful information and a long catalog of useful herbs listed alphabetically. When Culpeper gets to the dandelion, he dives right in, not mincing words: "DANDELYON, Vulgarly called, PISS-A-BEDS."[7]

Dandelion has long been known as a diuretic (a promoter of urination) and an extremely effective one. "It wonderfully openeth the Passages of the Urin both in yong and old,"[8] as Culpeper notes with enthusiasm. All the twentieth-century herbal reference books I consulted were in complete agreement with Dr. Culpeper, hailing the dandelion as an ideal diuretic.[9] Indeed, it can sometimes cause urination a little too strongly: to this day, when an English child plucks a dandelion, other children may tease him with a cry of "Pick a dandelion, wet the bed!"

The problem with other diuretic herbs — or even modern-day drugs — is that as they stimulate urination, they cause the body to lose valuable vitamins and minerals, especially potassium, a crucial nutrient for the heart and brain. But dandelion leaves contain an abundance of potassium, in a form that is easily assimilated by the body. So the dandelion's rich store of nutrients replaces what the urine washes away, and is particularly useful if the patient will be on long-term diuretics. Dandelion was, and is, used for treating conditions like pulmonary edema, gallbladder problems, kidney stones, gout — any condition where it's essential to get rid of excess fluid without depleting the body of nutrients it needs to thrive.[10]

Culpeper had a good deal more to say about the dandelion. "It is of an opening and clensing quality, and therfore very effectual for the Obstructions of the Liver, Gall, and Spleen, and the Diseases that arise from them, as the jaundice, & Hypocondriacal Passion . . ."[11] Again, modern texts bear him out; dandelion root has a well-validated ability to help the liver function at peak efficiency. Dandelions stimulate bile production and circulation throughout the liver, which acts as the body's filter and removes toxins from the blood.[12]

There seems to be no end to the dandelion's useful qualities. When used as a poultice for wounds, the leaves have mild antibiotic properties. Dandelion flowers have analgesic qualities, which means that they function as a pain reliever. Like Tylenol, they don't have any salicylates, the alkaloid compounds found in aspirin that can be irritating to the stomach, so an infusion of dandelion flowers works gently and subtly to relieve pain.[13] The mildly caustic sap from the stem and leaves can even remove warts.

Culpeper points out that dandelions are good for other, less familiar, conditions as well.

> It powerfully clenseth Aposthumes, and inward in the uritory passages, and by the drying and temperate quality doth afterward heal them . . . it helpeth also to procure rest and sleep to Bodies distempered by the Heat of Ague Fits, or otherwise. The distilled Water is effectual to drink in Pestilential Feavers, and to wash the Sores.[14]

Culpeper lists the herbs' qualities in the page margins, as a convenient reference for the harried care-giver. In the case of dandelions, he notes that they "Openeth, Clenseth," and are useful for "Obstructions, Liver, Gall, Spleen, Jaundice, Hypochodriacal Melancholly, Disury, Consumption, Cachexia, Watching, Heat, Agu, Pestilence . . ."[15] Everything from Pestilential Feavers to Warts — it actually is difficult to find a medical condition that dandelions *aren't* recommended for. One modern herbal lists its qualities as hepatic, cholagogue, lithotriptic, tonic, nutritive, galactagogue, stomachic, aperient, laxative, diuretic, deobstruent, bactericide, fungicide, astringent, and sedative.[16] It may be the perfect all-around medicine, a plant one should surely take along if one were bound for a desert island, or a New World. It's what we might call today a broad-spectrum medicine, one that can improve general health and therefore enhance the patient's ability to fight off many kinds of diseases. "And whoso is drawing towards a Consumption, or an il Disposition of the whol Body . . . by the use herof for some time together shal find a wonderful help."[17]

Nicholas Culpeper was a man on a mission. He passionately wanted to convince people that good medicine was to be found growing right under their noses.

> *You see here what Vertues this common Herb hath, and that's the reason you French and Dutch so often eat them in the Spring; and now if you look a little further you may see plainly without a pair of spectakles, that Forraign Physitians are not so selfish as ours are, but more communicative of the Vertues of Plants to People.*[18]

He wanted health care to be available to everyone, simple and understandable, and affordable. He had no patience with learned physicians who jealously guarded their knowledge, shrouding medicine in a veil of Latin and Greek incomprehensible to the layperson. Some doctors charged exorbitant prices, their motto "pay-or-perish"; Culpeper wasn't shy about condemning them: "Some men are so damnable proud and envious withal, that they would have no body know any thing but themselves; the one I hope will shortly learn better manners, and the other be a burden too heavy for the Earth long to bear."[19]

Culpeper insisted that "people need little other Physick than such Herbs as grew neer them."[20] But he was trying to buck a deeply ingrained prejudice. The idea that common roadside plants contained powerful medicine was a difficult one to accept. Most folks reasoned that whatever is rare and expensive must be worth the money; that which grows for free in every compost pile is useless. Dandelions were made valueless by their very accessibility.

The medical professionals tended to prefer other methods of healing, ones that didn't involve digging up roadside weeds. Some of their remedies were effective, some of them less than helpful, such as "heroic" doses of poisonous drugs like mercury or arsenic. Increasingly, there grew an unbridgeable dichotomy between the physician and the herbalist: the doctor and his drugs for the rich, the herb-wife and her herbs for the poor.

∾

Mid-November, a sharp cold that creeps into the bones. The air is dry, germs are spreading. I start to sniffle, then to cough; I don't think my problem is anything serious, it's just the cold and the shortening days, and general winter run-down-ness: a bad case of Hypochondriacal Melancholly, perhaps. My annual cold is inexorably approaching.

As Culpeper pointed out, ". . . truly my own body being sickly brought me easily into a capacitie to know that Health was the greatest of all Earthly Blessings, and truly he was never sick that doth not beleeve it."[21] He's right, you never appreciate being able to breathe through your nose until you have a cold. Every fall I feel the urge, like a migrating goose, to fly straight to the clinic and plead for an antibiotic.

However, after immersing myself in all this dandelion research, I decide to do some first-hand experimentation, and try the effects out for myself — to consult Dr. Experience, as Dr. Culpeper recommends. "I cannot build my faith upon authors words nor beleeve a thing because they say it, and could wish every bodie were of my mind in this . . ."[22] I go outside into the November chill to gather some dandelion greens.

There have been a few frosts already, and the grass is browned to hay, but the lawn still has some green patches. I inspect the maze of leaves and grass blades around my shoes. My yard is herbicide-free, so there's a healthy diversity of leaf shapes and sizes besides the grass. Clovers jostle plantains, chickweed rubs elbows with ground ivy, a tangle of grass stems weaves around them all.

A spark of yellow attracts my attention: by mid-November, the last of the wildflowers were long dead, I thought. But there's a scraggly, half-open dandelion blossom here, and another one there; a short-stemmed yellow flower surrounded by a rosette of jagged lion's teeth, the "leavs which alwaies abide green."[23] Dandelions are not only the first messengers of spring, they're the rear-guard of autumn — after flowering and seeding in early spring, they do it all over again in fall, the very last plant to bloom as well as one of the first. "It Flowreth in one place or other almost all the year long."[24]

I bend over, picking carefully, and a pale sun warms my back. I gather a few flowers — they're good, too, but it's the vitamin-rich greens I really want. The leaves are delicate and tender, the fresh new growth that springs up in the cool of autumn.

A window in the house behind me opens and my son leans out. "Lost your glasses again, huh? Need any help?"

"Oh, no, thanks," I call. "I'm fine."

He watches for a while, shaking his head. "Find what you're looking for?"

"Yep," I answer, and go on picking leaves. He shrugs and closes the window against the cold. The air is crisp, it's true, but as refreshing as a drink of spring-water. I enjoy the stretch and the sunshine, and the break from the computer. Dandelions really are amazing medicine. I feel better already.

Back inside, in the cluttered kitchen, I heat water on the stove, and when the pot is steaming I throw in a handful of leaves. Nothing dramatic happens; the leaves make a limp green huddle in the bottom of the pot, and the water slowly turns a pale yellow-brown.

I pour a cup, and sit, and sip. My palate is jaded by years of Starbuck's coffee, and dandelion tea tastes pretty much like hot water, just a faint bitter tang that puckers the mouth. A little honey would help. I close my eyes and let the warmth flow through me. The steam, the warmth, and the relaxation must be healing in themselves. I take another sip, and return to my research.

Dandelions, as *Taraxacum officinale,* were listed for centuries in the official pharmacopoeia, or listing of medicines and drugs, of England and many other countries. They were in the United States Pharmacopeia from 1831 until 1926, and were listed in the US National Formulary from 1881 until 1965.[25] The humble dandelion is as ubiquitous in herbals as it is on lawns, appearing in countless handbooks, texts, and guides to herbal medicine.

But dandelions, and other herbal remedies, fell from favor. The doctor, an important personage in a community, had no wish to go outside on a cold, sleety day and scratch around in the mud looking for dandelions. Even when doctors did use plant remedies, they tended to use preparations of the dried roots and leaves, not the fresh plant.

Vitamin C is an unstable compound; the process of cooking or drying makes it vanish. Vitamin A is best eaten fresh, and it doesn't dissolve in water-based infusions or decoctions. For dandelions' rich store of vitamins to do their work, the plant must be consumed while fresh. Many physicians, unsurprised when powdered dandelion leaves failed to be as effective as promised, noted with scorn that dandelions just didn't work.

The avalanche of powerful drugs in the post-World War II era swept dandelions and other herbs from the pages of the medical journals. Modern drugs and techniques have changed the nature of medicine just as the invention of gunpowder changed warfare. Dandelion Tonic today seems as antiquated as a bow and arrow would on a twentieth-century battlefield. Only a few die-hard herbalists still use dandelions as medicine, although the greens are making quite a come-back in salads . . .

Does the dandelion have any place in modern medicine? Culpeper realized that dandelions help the liver, the body's filter, remove toxins from the blood. I am increasingly aware that we are everywhere surrounded by toxins Culpeper never began to imagine. He wore clothing made of wool and his underwear was, doubtless, linen — he never heard of polyester, fabrics made from petroleum, or the chemicals used in permanent press. He sat on wooden chairs, walked on woolen rugs, but almost everything I wear, walk on or sit on, in these modern times, is made of synthetics. Even the chair I'm sitting on as I type is made of synthetic fabric and particleboard, held together by toxic glues.

I uneasily take a sip of coffee from a Styrofoam cup, and it occurs to me to wonder what that white foamy stuff is made of. A little research, and I discover it's a petroleum-based plastic made from styrene, which is classed as a possible human carcinogen by the EPA, and which can be transferred to food — the hotter or fattier the food, the more styrene it absorbs.[26]

It's endless, really. Is there a toy in Toys-R-Us that isn't made of plastic? Tap water is laced with chlorine. Countless items that we encounter in the course of the day — paint, computer VDU screens, nail polish remover — give off toxins like formaldehyde, xylene, toluene, acetone.[27] Every time I walk into a building and smell that chemically, weird "new car smell," I wonder exactly what it is that I'm breathing in.

So some tantalizing questions arise. What if dandelions could help strain some of this flood of toxins from our bodies? What if dandelions were not only good for ailments like the Hypochondriacal Passion, but had a role to play in fighting diseases perhaps caused by these toxins — diseases like, say, cancer? Dandelions are rich in a chemical called inulin, which is currently being much studied for its effect on immunostimulatory function.[28] Chinese traditional medicine has used dandelion as a treatment for breast cancer for over a thousand years. "Students would do themselves much good, and benefit themselves exceedingly in the Study of Physick if they would tak the pains to view the Vertues of the Herbs . . ."[29] Even if it's an antiquated weapon, a bow still has power when used by a skilled archer.

Nicholas Culpeper wrote hurriedly, complaining of the *"want of time."*[30] He finally finished his work, carefully noting on the last page that he had completed it at his house in Spitalfields, next door to the Red Lyon Inn. I like to think of him popping into the Red Lyon for a celebratory ale after he had corked up his ink bottle. He died less than a year later, the lingering effects of his wound aggravated by overwork; he was well-known for never refusing to treat the many patients who knocked on his door, and sometimes saw as many as forty sufferers in a day, using his beloved plants to bring health and comfort. He died after proudly remarking to his wife that "I did by all persons as I wish they should do by me . . . I never gave a patient two medicines when one would serve."[31] Dandelions, that all-purpose green first-aid kit, must have been often in his hands.

~

> I consulted with my two Brothers, Dr. REASON,
> and Dr. EXPERIENCE, by whose advice together with
> the help of Dr. DILLIGENCE, I at last obtained my
> desires, and being warned by Mr. HONESTY, a stranger
> in our daies to publish it to the World, I have done it.
> Nich. Culpeper.
> Spitalfields
> next doorto the red Lyon.
> Novemb. 6. 1652.[32]

Notes to Chapter 7

1. Culpeper, Nicholas. *The English Physician*. This version of Culpeper's classic is found in its entirety at http://www.med.yale.edu/library/historical/culpeper/intro.htm, the website of the Harvey Cushing/John Hay Whitney Medical Library of the Yale University School of Medicine. Many editions of Culpeper's book have been published: one of the most recent is *Culpeper's Complete Herbal*, Herefordshire, UK: Wordsworth Editions, 1995. *The English Physician* was first published in 1652.

2. Culpeper was not alone in writing down herbal lore — John Gerard and William Turner were two of several other herbalists of the time who recorded medical information on plants, but they didn't have Culpeper's wit, sarcasm, and passion that make his book such entertaining reading.

3. William Reeves, quoted on *http://gen.culpepper.com/archives/uk/places/wakehurst3.htm*, a UK-based website called "Culpeper Connections, The Culpeper Family History Site." Reeves was Culpeper's secretary, and wrote shortly after Culpeper's death.

4. *Ibid.*

5. *Ibid.*

6. Culpeper, *The English Physician.*

7. *Ibid.*

8. *Ibid.*

9. Millspaugh, Charles F. *American Medicinal Plants.* New York, NY: Dover Publications, Inc., 1974.

10. Tilford, Gregory. *From Earth to Herbalist.*

11. Culpeper, *The English Physician.*

12. Tilford, Gregory, and Mary Wulff-Tilford. *Herbs for Pets.* Irvine, CA: BowTie Press, 1999.

13. *Ibid.*

14. Culpeper, *The English Physician.*

15. *Ibid.*

16. Weed, *Healing Wise.* Woodstock, NY: Ash Tree Publishing, 1989.

17. Culpeper, *The English Physician.*

18. *Ibid.*

19. *Ibid.*

20. *Ibid.*

21. *Ibid.*

22. *Ibid.*

23. *Ibid.*

24. *Ibid.*

25. Weed, Susun, *Healing Wise.*

26. *http://www.epa.gov/opptintr/chemfact/styre-sd.txt,* the web-site of the United States Environmental Protection Agency, "A Fact Sheet on Styrene," December, 1994. It notes that "Urinary excretion is the major route of elimination of styrene" in the human body.

27. Wolverton, B.C. *How to Grow Fresh Air.* New York, NY: Penguin Books, 1996. Dr. Wolverton's fascinating and scary book lists over three dozen common household and office sources of chemical emissions such as formaldehyde, xylene, toluene, benzene, trichloroethylene, chloroform, and acetone, many of which are suspected or known carcinogens. The sources of emissions include synthetic materials used in carpeting, ceiling tiles, cosmetics, fabrics, facial tissues, grocery bags, nail polish remover, paper towels, photocopiers, and upholstery.

28. *http://www.ars-grin.gov,* a website of the US Department of Agriculture's Agricultural Research Service.

29. Culpeper, *The English Physician.*

30. *Ibid.*

31. *http://gen.culpepper.com/archives/uk/places/wakehurst3.htm.*

32. Culpeper, *The English Physician.*

Chapter 8

Bitter Herbs

Albany, New York
2005

I walk into the bustling grocery store, and begin to browse around the shelves stocked with food. At first glance it looks like your average grocery store: shelves with cans, boxes, bags; piles of produce; loaves of bread, bags of potato chips. But this is not just any grocery store, it's the Honest Weight Food Co-Op. Health food. The produce is organic, the bread whole-grain, the potato chips baked, not fried. I prowl up and down the narrow, crowded aisles to begin my search: I'm on the hunt for dandelions.

The produce counter: organic carrots and potatoes, papayas and melons. Sure enough, dandelion greens are in the salad section, and I snitch one; a small, tender leaf with a bitter tang. $5.99 a pound.

There's dandelion extract in the pharmacy section, in a drug called Liver-Cleanse. And tincture of dandelion, $9.99 an ounce. Then the beverage section: freshly ground organic dandelion root for a no-caffeine coffee substitute. I pick up a package and inspect it; it doesn't look like ground coffee, it's a twist-tied plastic bag full of little tan-and-brown chunks. I open the bag and steal a sniff: not the robust smell of roasted coffee, but a nice, faint scent, reminiscent of earth and *autumn leaves. I glance at the price tag, and my jaw drops. $31.75 a pound.*

Dandelions are the most expensive item in the store. They're pricier than dried rose petals. Pound for pound, dandelions out-price prime rib, swordfish, and lobster.

Long ago, when I was a whippersnapper working my first job at an environmental education center, I had rather a crush on another employee, a cute young fellow with warm brown eyes and a slightly raffish beard. He was also a wonderful cook (unlike me) and he enjoyed spending time dabbling in the kitchen; he was one of those intuitive cooks who ignore the recipe books and throw in a pinch of ginger and a hint of oregano, and it all comes out tasting great. Part Polish, part Italian, he could cook a mean piroshki or lasagna. Being a fellow nature buff, he even experimented with wild edible plants, and he was especially fond of dandelions.

He assured me that dandelions were delicious, I was flirtatiously skeptical, and he vowed to convince me. One evening, I think it was at a staff party, he made a special treat: dandelion fritters. There they sat on a paper plate, golden balls of dough about the size of doughnut holes, deep-fried in crackling oil. "Here, try this," he said, holding one out on the end of a fork, and smiling. (He really had very nice eyes. I later sampled fried grasshoppers when he asked me to.)

"Wow!" I said, chewing. "Delicious!" I wasn't stooping to flattery, it was delicious, rich and spicy. He gave me that dazzling smile. "However do you make it?" I inquired. "I never would have guessed that dandelions would taste so good."

He smiled, casually tossing another fritter into the hot oil, where it bounced with a vicious sizzle. "Well, let's see," he said, leaning back on the kitchen counter. "You take some flour, and add about two cups of grated cheese — extra sharp cheddar. Then you add the spices, some powdered mustard, horseradish, and a cup of onions sautéed in butter, and two teaspoons of chopped dandelion flowers."

I had another one — they really were tasty. "And dandelions are so good for you, too," he added with enthusiasm. I glanced at the sizzling oil in the deep-frier, but quickly averted my eyes. Love is blind. "See what I mean about wild edible foods?" he asked, and I replied in the affirmative without hesitation.

However, the romance didn't work out, either with the cute young chef or with the dandelion. Because, of course, if you made the exact same fritter recipe and forgot the dandelions, it would taste precisely the same. In their more unvarnished state, dandelions are . . . well, let's just say that they'll never replace filet mignon. They're not bad, mind you. They're just not all that good.

Oh, the leaves are a tasty salad green, for those of you who like salad greens. I eat salad more out of a sense of duty than a desire for gastronomic pleasure. If the Surgeon-General were to determine that salads were hazardous to my health, they would be an easy habit to break. Of course, you don't have to eat the greens raw; cookbooks assure you that dandelion leaves can be lightly boiled, like spinach, and they taste amazingly like . . . boiled spinach. I will say no more.

Or the leaves can be infused, which means steeped in warm water, and made into a tea. The water should be warm but not boiling, so as to retain the maximum amount of nutrients. The resulting liquid is richly nutritious and has many medicinal qualities, but that's another chapter. It's medicine, it's not really tasty. Dandelion tea, famed for its diuretic properties, is a pale yellow watery drink that looks alarmingly like . . . well.

Then I tried what the books call Dandelion Coffee. It should come as no surprise that dandelion coffee is not Starbuck's Mellow Roast. But let's be fair, that's because a dandelion root isn't a coffee bean. It's something completely different, and to compare ground dandelion root to coffee beans is like comparing it to hamburger, or blueberries, or champagne. They're completely different things. You may like dandelion coffee, you may not, but you have to get the word "coffee" out of your head when you drink it. What's harder to get out of your head is the price tag, approximately ten times the cost of Maxwell House.

Of course, you don't have to buy dandelion roots. You can just go out and dig them — they're certainly not hard to find. The wild foods cookbooks airily urge you to dig up dandelion roots, but, like much advice, it's easier said than done. It takes a fair amount of elbow grease to unearth even one root, yank it free of the web of clinging rootlets, and shake the dirt off. It's recommended to look for roots that are neither too old nor too young; the midsize ones have the best flavor.[1]

But once the dandelion roots are dug, then the hard work starts in earnest. The wild edible food books all agree that you need to wash the roots, then dry them in a low oven, then grate or grind them up. Again, easy to say. Once you've tried to wash every speck of dirt out of the gnarly crevices of a dandelion root, you begin to understand why the stuff sells for $31.75 a pound. Then they have to be ground up. It's all a lot of good healthy exercise, no doubt; those who gather their food from the wild rarely get fat doing it. I sip a sour cup of dandelion coffee and survey my dishpan hands after scrubbing a dozen dandelion roots, and sigh. A trip to Starbuck's is a lot less work . . .

Still, I find myself pondering dandelions, even as I unwrap a chocolate truffle to get the bitter taste out of my mouth. Dandelions aren't bad, I suppose. With the right trimmings, they can taste pretty fair. But the bottom line is that they're really, really good for you, a complete food that contains protein and carbohydrates as well as hefty amounts of vitamins. Low in calories, too.[2] I should eat more greens, I know. It's just that . . . dandelions don't taste all that good. Too bitter.

"And they shall eat the flesh in that night, roast with fire, and unleavened bread; and with bitter herbs they shall eat it."[3]

The grim command is in the Old Testament, recorded thousands of years ago. The stern God of Exodus required of the Hebrews who were enslaved in Egypt that they should eat the flesh of a roasted lamb with bitter herbs, and He made it sound like a punishment. To this day, bitter herbs, often including dandelion greens, are a traditional part of the Passover Seder, to remind participants of the bitterness of that captivity.[4] But it's not a punishment, it's a very good idea.

Why is it wise to eat bitter herbs? There is value in bitterness; a difficult lesson to learn. Just because something's bitter doesn't mean it's bad for us. Sweet is what humans crave, from the first sweet taste of breast milk; we are born with a sweet tooth. It seems to be against our pleasure-loving nature to eat that which is bitter. So as early humans developed agriculture, at harvest-time they carefully chose which seeds to sow in spring, and they selected the tenderest, juiciest, and above all the sweetest fruits and grains. To this day, when horticulturists are deciding what varieties of tomatoes or lettuce to propagate, they tend to throw away the bitter and choose the sweet.

But the worldly-wise prophets of the Old Testament had seen a thing or two, and they knew that life isn't always sweet. The bitterness in plants is caused by phytochemicals that are richly health-giving.[5] Iceberg lettuce is sweet, or rather, bland — and it's blandness that is the direct opposite of bitterness. My grandmother, used to home-garden-raised vegetables, used to complain loudly about grocery store lettuce and pale pink tomatoes that tasted like the squares of Styrofoam they sat on. "Plastic vegetables," she'd grumble. "You'd get more vitamins from eating the wrapping."

And — as is traditional during the Seder — dandelion greens and other bitter herbs are best eaten just before the main meal. Only a few bites of dandelion greens trigger the production of bile and other digestive agents — they prepare the soil, so to speak, gently awaken the digestive system and get it juiced up and ready to go. The body can digest food more efficiently, and therefore absorb more nutrients, and the liver can more effectively eliminate toxins.[6]

So dandelions are good for you, and therefore a certain sense of virtue attends the eating of them. My stern Protestant grandmother always made me finish my vegetables, and frowned on dessert. But what of the more frivolous part of our nature, the sweet tooth we were born with? Do dandelions have no sweetness for us to taste?

~

Dandelion wine. No need to brave the cold of March to dig bitter roots and greens from beneath the snow. To make dandelion wine, you need the full-blown blossoms. What's more, you have to harvest the flowers when they're open.

Since dandelion flowers close up tight on cloudy or rainy days, you have no choice but to wait for sunshine. All recipes for dandelion wine agree that you have to pick the flowers in full sun, preferably at noon when they are open wide. It's important to avoid roadsides or lawns where car emissions or herbicides may pollute the flowers.[7]

So you have to find a noon-time summer meadow, full of bright yellow flowers, and pick them with the warm sun shining down. It almost makes you feel like twirling with your arms outstretched, singing "The Sound of Music." It beats a trip to the grocery store, anyway.

Pick the warm, sweet-smelling blossoms, boil them with sugar and orange peel, and let the mixture ferment. Pour the pale liquid into sparkling glass bottles, and put it in the cellar till winter. Traditionally, dandelion wine was saved to be drunk at the winter solstice.[8]

The wine is well worth the effort. It's light yellow, faintly sweet, a little sparkly, like champagne. Have a glass, drink in the warmth of the summer meadow to fight off the chill of January. Dandelion wine has long been a traditional cure for the illnesses of winter: colds, flu, coughs, the midwinter blues in general. It's very pleasant medicine. Have another glass!

Ah, but remember the advice, or warning, or prophecy, of the Old Testament: "They shall eat bitter herbs." Bitterness: there's no escaping it, in the end. What's the price of overindulgence in wine?

The unpleasant symptoms of the morning after are the result of a build-up of toxins in the body. In order to recover from a hangover, the liver has to filter the toxins out of the bloodstream, and then they must be flushed out of the body. Dehydration is one of the causes of the well-known hangover headache. And (all together, now, class) what herbal remedy can stimulate liver function, and also serves as a diuretic, flushing toxins from the body while replacing the nutrients and fluids that the body needs?

I think I'll have a cup of dandelion tea, to repair the effects of over-consumption of dandelion wine. I will sip the soothing drink, and reflect on vice and virtue, bitter and sweet, all bound up together in one humble roadside plant.

$\approx$

Recipes

I'm no chef, as my friends and family would be quick to tell you, so recipes are not my forté. But fortunately the versatile dandelion doesn't require much culinary expertise.

Flowers

All parts of the dandelion plant are edible, in the sense that they're not poisonous, but some parts taste better than others. The stem is pretty much useless, being very bitter and sticky, and the bracts and sepals — the greenish bits that surround the flower head — are bitter, too. I've actually had the best luck with the flowers — what most people call the "petals." They have a bland, faintly sweet flavor, but their color brightens up a meal.

Gather the flowers on a sunny day, wash off any insect visitors, and pull or snip the yellow florets free of the green sepals and stalks. You can also use the blossoms of violets, nasturtiums, the ubiquitous mints, such as ground ivy, or red clovers — just add them to recipes or sprinkle them on salads or cake frosting or whatever, as colorful decorations for any dish.

Leaves

Gather the leaves from plants that have not yet bloomed. This makes them a lot harder to find, but after the plant flowers, it undergoes a chemical change, as lettuce does when it flowers, or "bolts." This makes the leaves much more bitter. Spring dandelion greens are famous, but there's a period in fall when the dandelion re-leafs, and has fresh, tender leaves. After a frost, they're even sweeter. You can use dandelion leaves in any recipe that calls for spinach.

If you're the sort that plans ahead, putting a bushel basket or something to shade a clump of dandelions for a few weeks will "blanch" the leaves and buds — they turn white as the plant stops photosynthesizing in the absence of sunlight. This removes the bitter taste.

Dandelion Quiche

5 eggs
1/2 cup milk (low-fat is good)
3/4 cup grated cheese
Dandelion leaves and flowers — how many you use depends on how daring you are. Start with a small handful.

Lightly sauté the dandelion leaves and flowers in a teaspoon of oil or butter. Sprinkle with salt and pepper. Add milk to eggs, and beat with a fork till foamy. Add dandelions and grated cheese. Pour into a baked pie crust. Sprinkle with dandelion flowers. Bake for 30 minutes at 350°.

Dandelion Tea

Gather as many dandelion leaves as you need for immediate use — about a dozen leaves make two cups of tea, but gather more or less depending on how strong you like your tea. Wash the leaves and add to hot (not boiling) water.

Let steep approximately fifteen minutes. Add honey, sugar, or even a teabag of regular tea for flavor.

A wonderful invention called a "French press" can be used to make small amounts of herbal teas. It's a sort of coffee pot with a strainer. Place the leaves in the bottom of the pot, add hot water, and let it steep; then push down a lever and it will press the leaves down to the bottom of the pot so you can pour out the clear liquid through the strainer.

You can add any edible plant to this process to flavor the tea to your own taste. Throw in a handful of pine needles, or violet greens, or red clover flowers, or strawberries, to create a unique tea.

Some cautions

• To harvest dandelions, be certain you're not gathering them from a pesticide-treated lawn, or anywhere near a roadway. Some sources recommend staying as much as 100 feet from a road to avoid lead and other emissions from cars.

• As with any new food, let your system get used to them gradually, in small amounts.

• Remember not to drink too much dandelion tea late in the evening. They're called "piss-a-beds" for a reason.

• Always be sure you have properly identified any wild plants before eating them. That said, according to the *Field Guide to North American Wild Plants* by Peter Dykeman and Thomas Elias, dandelions have no poisonous look-alikes. Some plants, like Queen Anne's lace (also known as wild carrot) are edible, but have poisonous look-alikes — deadly plants that look so much like them that it's easy to confuse the two. Queen Anne's lace closely resembles its cousin, the highly toxic poison hemlock (the stuff that Socrates drank.)

• Always consult a physician before taking any plant for medical reasons. Also, if you are pregnant or have any condition where a diuretic would be contraindicated, use dandelion sparingly.

Notes to Chapter 8

1. Brown, Tom. *Guide to Wild Edible and Medicinal Plants.* New York, NY: Berkley Books, 1985.

2. Gibbons, *Stalking the Healthful Herbs.*

Protein	Calcium	Iron	Potassium	Vitamin A	Vitamin C
dandelion greens (100 grams)					
2.7 g	187 mg	3.1 mg	397 mg	14000 I.U.	35 mg
ice berg lettuce (100 grams)					
0.9 g	20 mg	0.5 mg	175 mg	330 I.U.	6 mg
spinach (100 grams)					
3.2 g	93 mg	3.1 mg	470 mg	8100 I.U.	51 mg
oranges (100 grams)					
1.0 g	41 mg	0.4 mg	200 mg	200 I.U.	50 mg

3. Exodus 12:8, *The Holy Bible,* King James Version.

4. Walker, Winifred. *All the Plants of the Bible.* Garden City, NY: Doubleday and Co., Inc., 1979. The Bible doesn't, of course, actually mention dandelions. The text of Exodus 12:8 just says "bitter herbs," without specifying any particular plants. Students of the Torah have generally identified dandelion as one of the appropriate bitter herbs, with endive, chicory, lettuce, sorrel, watercress, romaine, and horseradish as other possible candidates.

5. Duke, James, *Medical Botany,* available at *http://www.ars-grin.gov/duke/syllabus/module6.htm.*

6. Tilford, *Herbs for Pets.*

7. Richardson, Joan. *Wild Edible Plants of New England.* Yarmouth, ME: DeLorme Publishing Co., 1981. Richardson recommends keeping fifty feet away from a well-used road or highway to avoid gathering plants that have been polluted by car exhaust and road construction. *The Mother Earth News,* issue 15, May/June, 1972, *The Plowboy Interview: Euell Gibbons,* recommends that collectors gather plants fifty or even a hundred feet from the road to avoid contaminants, especially lead. Available

at *http://www.motherearthnews.com/top_articles/1972_May_June/ The_Plowboy_Interview.*

8. Recipes for dandelion wine abound, in cookbooks and on the Internet. Several nice ones are at http://www.prodigalgardens.info/ may%20weblog.htmrecipe. Also see *http://lost-heritage.com/ boutique.html* or *http://www.mistrettas.com/winenov.html* for ordering dandelion wine.

Chapter 9

The Devil's Invention: Are Dandelions Bad?

Amsterdam, New York
2006

A spring day, a little overcast and chilly outside, but inside the greenhouse the temperature is tropical. The vast array of garden plants for sale spreads out in front of Timothy and me like a vast quilt, in polyester-bright hues of scarlet, hot pink, magenta; the wild sweep of colors is much to the taste of an eight-year-old who wants his entire bedroom painted orange to go with his purple bedspread. He gazes around with enthusiasm, then suddenly wrinkles his small nose.

"Hey, Mom, what's this?" he demands, bending down to peer under one of the shelves of petunia flats. I follow his pointing finger to see a wild sprawl of greenery writhing across the neatly-swept concrete floor, a mass of broad, jagged leaves each about two feet long, reaching out like octopus tentacles.

"It's a dandelion," I said.

"You're crazy, it's too big," he said. But it was. The dandelion seed had sprouted in a crack in the concrete floor, as dandelions tend to do. But this particular seedling was

well-watered by the drippings of the heavily-fertilized plants
above, and had reached a sort of dandelion nirvana. The plant
seemed to be growing almost visibly; I expected yellow flow-
ers to start rocketing up under our noses like firecrackers.

The greenhouse owner stalked by and spotted the in-
truder, and her eyes opened wide in horror. She promptly bent,
grabbed the leaves, and started pulling. The dandelion put
up quite a battle, but she won at last. She stamped off, car-
rying a huge armful of already-wilting leaves.

I felt a little sad at the bold dandelion's demise. It
seemed unfair to extinguish a living thing that was so full of
vigor. And I was aware of a sense of admiration for the plant
— the grudging admiration that's earned by any successful
thief, any cheeky flaunter of society's rules, a Robin Hood
or a Butch Cassidy. The dandelion, so successful at sneak-
ing into jealously guarded lawns and gardens, has for me a
little of the romantic allure of the gypsy, the rebel, the outlaw.

Robin Hood gave back to his society, though; he may
have stolen from the rich, who could well afford it, but at
least he gave a percentage of the profits to the poor. Dandelions, surely, are more like Butch and Sundance, thieves who
didn't worry about sharing their stolen riches but kept all
the good stuff for themselves. Or are they?

Later on that chilly spring day, we went for a stroll in
the town park, Timothy and I, and we spotted a butterfly. It
was a gorgeous creature, one of those big showy ones that
seem to belong to a tropical rainforest, with golden streaks
on coal-black wings. It was following an erratic flight-path,
up, down, back and forth, as though in search of something.

"I wonder what he's looking for," Timothy remarked,
frowning.

"Maybe he's hungry," I said.

We glanced around the tidily-groomed park. There were no flowers in the squared-off beds, not yet, only the hard, round buds of tulips, clenched like fists. The butterfly searched high and low, veering from the short-clipped grass up into the tree-tops, but there was nothing to be found but greenery. Finally, it zigzagged under a hedge, where there lurked a lone dandelion that had escaped the mower.

The butterfly landed on the yellow disk, which tilted and bobbed under its weight. The magnificent wings fanned slowly, as the butterfly uncoiled a long thread of tongue and inserted it with accuracy into each of the dandelion's florets, siphoning the nectar from them one by one. It was a long time before the butterfly flew away.

Dandelions give rich nutrition to wildlife, the nectar, seeds, and leaves used by dozens of species. That's true of many plants, of course, but dandelions are uniquely useful, since they are among the very earliest of spring flowers to leaf out and bloom. They provide food in habitats — especially in human-altered places — where there *is* no other food. In places like the well-groomed park where we watched the butterfly, it's sometimes a choice between dandelions or starvation.[1] January and February are the only months in which I have failed to find a dandelion in bloom — and that's in chilly upstate New York.

Dandelions bloom early, and seed early, but they have a major period of re-bloom in fall, a thick harvest of new tender leaves and nectar-laden flowers that quickly seed. I've seen a flock of yellow and black goldfinches, like a swarm of giant bees, hopping over an autumn-dry park lawn and feasting on dandelion puffballs. This late blooming provides crucial nutrition for animals who are storing up energy for the most desperate challenges of the year: the grueling fall migration, or the long winter sleep. Dandelions are both the appetizer and the dessert course of nature's bounty; they're available to wildlife, as to the scurvy-ridden settlers, precisely at the seasons when they are most desperately needed. [2]

Dandelions are good, right?

~

Whenever Timothy watches a movie, or listens to a story, he always has one question about each and every character. "Is that a good guy?" he demands. Is he (or she) good? Or bad? It's that simple, and he wants no character analysis, no long-winded exploration of motives. There's two sides only: Harry Potter or Lord Voldemort, Robin Hood or the Sheriff of Nottingham, angel or devil. Good or Bad.

On this simple scale of values, where do the dandelions fall? Are they Luke Skywalker or the Evil Emperor? Well, they feed lovely butterflies. They're good medicine, and nutritious food. They infest lawns, but that's purely a cosmetic issue. If I don't care if there are dandelions in my yard, should I care if there are dandelions in yours?

Is there a problem with dandelions having spread all over the country — all over the world? In this nation of immigrants, it seems as though the plants that the settlers brought with them should be as welcome as the settlers themselves. From the environmental point of view, are dandelions bad?

Well, of course it's not as simple as that. Ecologically speaking, it's not a question of good or bad. Spiders and snapping turtles aren't "bad." Butterflies aren't "good." The burning question is: are they native to their environment, or nonnative? Whether you love dandelions or loathe them, the fact remains that, away from their native Eurasian soil, dandelions are alien. They're not a part of the original landscape of the Americas, or Australia, or Africa, or any of the innumerable islands they have spread to. They're not connected to the intricate network of relationships that has evolved in these places over millennia. Alien plants are like E.T., the Extra-Terrestrial, out of place and far from home.

But unlike the cuddly E.T., plants are shockingly aggressive organisms. We tend to think of the denizens of the Plant Kingdom as passive victims, submissively waiting for a cow to come and chew on them. But plants can be as brutal as a horde of invading Mongols. Plants beat back their enemies with spikes and thorns, poison their competitors, strangle their hosts. Some plants have Machiavellian strategies for survival: they hide, they mimic more deadly plants, or reproduce faster than cockroaches. Every species of plant is engaged in a perpetual struggle to get as many of its genes into the pool as possible, to carpet the planet with its own kind. But in its native environment, the attempt to take over the earth is continually thwarted by a host of competitors and predators.

The human colonizers who spread so many plants worldwide didn't stop to consider this hard, cold ecological truth. The Europeans brought dandelions to the New World, but what they forgot to pack were the subtle natural mechanisms that could keep dandelions in check. There's an intricate combination of predators, fungi, bacteria and other organisms that have co-evolved with dandelions, all interwoven and bound together through the slow ages of time, as well as the effects of weather, temperature, shade, erosion, and myriad other factors. But when dandelions were escorted across continents and oceans, these restraints were left behind. The nibbles of goldfinches and the sippings of stray butterflies don't even make a dent in the dandelion population.

So back to the question, good or bad? How much of an environmental problem is the intrusion of an alien species into an ecosystem?

Well, as invaders, dandelions are relatively benign. Dandelions didn't invade virgin prairie; they could never have elbowed aside native plants like the big bluestem grass that grew as high as a man's head. Humans had to open a pathway for them first. Dandelions didn't lead the way into the West — they followed behind the wagons and bloomed in the wheel ruts. They didn't invade the dense forests of New England, either — they bided their time in the herb gardens till the settlers cut down the trees.

To this day, dandelions don't pose a threat to wilderness areas. Oh, some hardy individuals might venture a few yards into the forest, or pop up where a storm has toppled an old oak, opening a skylight in the canopy of leaves. But dandelions, like most seral species, tend to stick to the open — they linger in the parking lot like timid urbanites, and don't venture onto the back-country trails. Dandelion seeds aren't adapted to hitchhike on the fur of a passing squirrel or raccoon — they need wind. On the forest floor, there's little wind to disperse the seeds. And the sun-loving dandelion always perishes in the shade.

Other non-native plants are not so harmless. Oriental bittersweet is a beautiful vine that wraps itself around a young oak or a sugar maple and slowly strangles it. Cheatgrass, a spiky alien pest that's all but barren of nutrition, has carpeted a hundred million acres of former prairie. Water hyacinths creep insidiously over a pond: one year there's a few of the dainty lilypads speckled over the surface of the water, and everyone remarks on how pretty they are. Next year there are a few more. Five years later, you can practically walk on the dense mat of green that deprives the pond of oxygen like a plastic bag over a child's face. Kudzu, purple loosestrife, water chestnut: invasive plants infest habitats around the world. The list of aggressive alien plants is getting longer year by year, and the problem of invasive species is considered to be one of the worst environmental hazards of the twenty-first century.[3]

But dandelions won't be strangling any sugar maples, they drown in wetlands, and can't compete with tall native grasses. Dandelions need the soil to be disturbed, and the shade to be removed. That's why dandelions seem to go looking for human activities: roadsides, construction sites, parking lots — and lawns. Having escaped the herb gardens a few decades ago, they now seem on a quest to get back into the yards they once abandoned.

But — the question returns with the ruthless persistence of an eight-year-old — are dandelions bad? If forced to answer the question, I'd say that dandelions are a bad thing, an alien species, with many redeeming virtues. They're an unbelievably nutritious plant that heals the earth as well as people, they're medicine, and magic, and, yes, beauty . . . Oh, and did I mention that they tend to move into sunny areas of mowed grass? But in the great scheme of the world's ecology, what's a few dandelions on a lawn?

Well, some homeowners don't like weeds, of course. And, I must admit, they have something of a point. Dandelions interrupt the smooth expanse of green that's so pleasing to the eye, and after the seeds fly off, the stems are unsightly. Some irate greens keepers go so far as to describe dandelions as an infestation, a curse or a disease. And it's true that dandelions don't really belong on the North American continent. They never will. They'll always be outlaws.

But if dandelions are a disease, they're surely a minor one, compared to the creeping cancer of invasive species that threaten to eradicate whole ecosystems. Dandelions are an ailment like the common cold, wide-spread but relatively harmless. But the cure for the dandelion "disease" has led to one of the greatest environmental disasters of all time.

Notes to Chapter 9

1. Morse, Roger A. *The Complete Guide to Beekeeping*. New York, NY: E. P. Dutton, 1972.

2. Martin, Alexander C., Arnold L. Nelson, and Herbert S. Zim. *American Wildlife and Plants: A Guide to Wildlife Food Habits*. New York, NY: Dover Publications, Inc., 1951.

This fascinating and exhaustive work is based on the analysis of the stomach contents of thousands of animals. Dandelions are listed as food for more than thirty species. The pocket gopher's diet, based on a study in Utah, is 67% dandelion seeds and foliage.

3. Manning, Richard. *Grasslands: The History, Biology, Politics, and Promise of the American Prairie*. New York, NY: Penguin Books, 1997.

Chapter 10

The Devil's Arsenal

Silver Spring, Maryland
1961

A small, thin-faced woman sits at a typewriter in an office crammed with books, papers, and scientific journals. She is typing rapidly, and the keys' clatter fills the quiet, sunny room. She is writing a book, and the message she has to tell is terrifying. She knows she has not much time in which to tell it.

But after hours of work, she takes a break. It is springtime, and she loves the spring, especially the sweet sound of birdsong. She is too ill to go birding these days, but she can still enjoy birds from the window of her suburban Maryland home. She glances out the window and smiles at the sight of a robin in a neighbor's yard.

The robin stands on the green square of neatly-trimmed grass, not a dandelion in sight. He takes a step, and then tilts his head, listening, as his bright black bead of an eye stares at the ground. He hears a tiny sound, and tenses, ready for the kill. Abruptly, his beak flashes down and snares a fat worm, dragging it wriggling from the soil. His red chest puffs with pride, and he spreads his wings and flies home to where a nestful of young robins wait, hungry mouths gaping wide.

The woman smiles as the robin flies away, her dark eyes lightening; then she sighs and bends over the typewriter again. Rachel Carson loves birds, and spring — and she has work to do.

It's the very definition of springtime — the cheerful sight of a robin pulling a worm out of the new green grass. But the robin who goes worm-hunting on a well-tended lawn may be bringing more than a tasty meal home to his youngsters. If the neatly-trimmed, dandelion-free lawn has been treated with a pesticide, the robin and his family may become statistics. It's estimated that more than seven million wild birds are killed by the aesthetic use of lawn pesticides in the United States, every year.[1] Aesthetic use: that doesn't mean agricultural pesticides to grow food crops. That means the use of pesticides to make our lawns and gardens look nice.

Seven million wild birds, dead. To get rid of, among other things, dandelions.

Rachel Carson's classic book *Silent Spring* woke the world to the dangers of pesticides. I have a copy of it, an old 1964 paperback I bought for a quarter at a garage sale. "The explosive best-seller that is shocking the world!" screams the advertising on the cover. But I rarely encounter anyone, these days, who has actually read the book. No wonder, really — *Silent Spring* may have been an "explosive best-seller," but it's not exactly a thriller. It's long and dull, full of depressing details and statistics, very tough going. Knowing that the pesticide industries would oppose her findings bitterly, Rachel Carson spent years in painstaking research, in order to be sure that the facts her book contained would be incontrovertible, and her grim findings don't make for light reading.

I keep the old book on my desk, though, and every now and then I pick it up. I leaf through it, skimming over statistics on fish kills and neurological disorders, and come across a mention of springtime robins — "numbers of doomed birds . . . in the agonized tremors that precede death."[2] I close the book hastily. I don't want to read any more.

Really, I think, Rachel Carson didn't want to write the book any more than I want to read it. She would have preferred to spend her time enjoying the wonders of nature and writing about the beauties of living birds, not dying ones. She at first tried to persuade others to take on the pesticide battle, offering to supply them with material.[3] But no one else was willing to embark on such a dismal and controversial project. Finally, she wrote that "I could never again listen happily to a thrush song if I had not done all I could."[4]

Unlike *Silent Spring*, the label of a container of chemical pesticide — any brand — is certainly not dull reading. All sorts of words usually found in murder mysteries and spy novels leap out at you. "Fatal . . . poison . . . danger . . . death . . ." Anyone, of any age, can buy pesticides off the shelf and use them at home without any sort of license, and therefore we tend to assume that lawn and garden pesticides are trivial, everyday stuff, as harmless as dish detergent — just to clean up your lawn a trifle.

What if a bird-lover puts up a feeder on a lawn that has been treated with herbicide for dandelions? People enjoy seeing sparrows and chickadees hopping around on the grass, pecking at seeds. But they don't see what happens after the birds fly away. Our knowledge of pesticides' effects on birds is like a pyramid-shaped iceberg, in which the widest portion by far remains unseen. Only a few dead birds are noticed and collected, but it's probable that the majority of bird deaths caused by pesticides go undetected.[5]

Pesticides are poisons. They kill living things. On every package there is the number of a poison control center, or instructions to physicians. Obviously, if you drink the pesticide, it's a threat to your health. But even if you follow the directions and apply the pesticide precisely as instructed by the manufacturer, pesticides kill things.[6] Dandelions, but also robins. Weeds, but also butterflies, cardinals, song sparrows, chickadees, dogs, cats, ladybugs, bluebirds, meadowlarks . . .

~

And children? Children have long been identified as being especially vulnerable to pesticide exposure. Their developing bodies are at greater risk from toxins.[7] Also, they don't sit in chairs on the deck, sip iced tea, and merely gaze at the lawn. Children play on lawns: they roll down the hill, lie on the grass, chew grass blades, blow dandelion puffballs. Like the robins and the neighborhood dogs and cats, they can't read the little signs with the tiny letters warning that this lawn has been treated with herbicides.

Rachel Carson waded though enormous piles of reports and scientific documents as she quoted the horrifying statistics of wildlife and human deaths caused by pesticides — I've counted more than five hundred end-notes in *Silent Spring*. A true scientist, she wrote for the most part in a clinical, almost detached style as she reported the details of bird kills. But sometimes her anger broke through, and one can feel her passion: "Who has made the decision that sets in motion these chains of poisonings, this ever-widening wave of death that spreads out, like ripples when a pebble is dropped into a still pond? Who has placed in one pan of the scales . . . the pitiful heaps of many-hued feathers, the lifeless remains of the birds . . . ?"[8]

Finally, after years of work and struggle with ill heath, the book was finished. "Last night the thoughts of all the birds and other creatures and all the loveliness that is in nature came to me with such a surge of deep happiness, that now I had done what I could — I had been able to complete it — now it had its own life . . ."[9]

When *Silent Spring* was published, it caused a sensation. Rachel Carson was both applauded and lambasted, praised as a heroine and vilified as a hysterical woman. But the book really changed things. The deadly chemical known as DDT was banned in the United States, and the public was made aware of the dangers of pesticides. Rachel Carson was showered with medals, honors, awards, and credited with starting the environmental movement. But she was a shrewd lady. "It would be unrealistic to believe one book could bring a complete change," she wrote.[10] Rachel Carson died of cancer, less than two years after the publication of *Silent Spring*.

～

I sit at the breakfast table, sip coffee (Starbuck's, not dandelion) and open my morning newspaper. *Pesticides Infiltrate Water* is the headline, in letters an inch high. Guess Rachel Carson was right about not having licked the problem yet. *Silent Spring* jolted us awake like an alarm clock, but it's all too fatally easy to press the snooze button. Forty years after the publication of *Silent Spring*, the average American homeowner uses ten times as much pesticide per acre as the average American farmer; an arsenal of deadly chemical weapons is unleashed, every day, on pleasant, green lawns.[11]

I'm tempted to turn the page, and read Dear Abby or the sports section instead of the pesticide article. But I sigh, take another sip of coffee, and begin to wade through the dense columns of print. "Almost all of the nation's rivers and streams — and the fish in them — are contaminated with pesticides linked to cancer, birth defects, and neurological disorders . . ." I try to think of an excuse, something I should be doing, so I won't have to continue reading. But I heave another sigh and plunge back in. "Pesticides were found in rivers and streams between 1992 and 2001, says a study released Friday [March 2, 2006] by the US Geological Survey . . . Forty pesticides had a widespread presence in streams and sediment in both urban and agricultural areas, at concentrations that could affect aquatic life or fish-eating wildlife . . ."[12] I really don't want to read this. But I have to — we all do. We need to know what's in our water, our food, our air — and on our lawns.

~

A lawn that has been drenched with pesticide looks like any other lawn, pretty much. Pesticides are more or less invisible: they have no strong smell, there's little sign of their presence. How can I tell which lawn I can let my child play on? Where can a robin get a worm to bring home to the nestlings? What's a parent to do?

Seems like there should be some sort of marker, a badge of safety, a brightly-colored seal of approval on the grass of organic lawns. Something easy to spot, scattered across the grass to tell children that this field is safe to play on, this hill is safe to roll down. Some signpost should promise that the worms that a robin pulls out of this lawn won't bring slow death to baby robins; the dog who plays catch on this grass won't be at risk for neurological damage; walking barefoot on this lawn won't bring a higher risk of leukemia to my toddler.

Well, there *is* such a marker, and it's bright yellow. I like to imagine Rachel Carson smiling, as she gazes at the birds on my lawn and the grass sparkling with dandelions.

Notes to Chapter 10

1. Wargo, John. *Risks From Lawn Care Pesticides.*

2. Carson, Rachel. *Silent Spring.* Boston, MA: Houghton Mifflin, 1962.

3. Brooks, Paul. *The House of Life: Rachel Carson at Work.* Boston, MA: Houghton Mifflin, 1972.

4. Rachel Carson to Dorothy Freeman, January 23, 1962, as quoted in Brooks, *The House of Life; Rachel Carson at Work.*

5. Rachel Carson Council News, No. 91, spring 1999. The Rachel Carson Council, Inc. is a not-for-profit organization that promotes environmentally benign pest management strategies.

6. *Lawn Pesticides, An Unacceptable Risk,* a Publication of Grassroots Environmental Education, 2003. Grassroots, Inc. is a not-for-profit organization dedicated to environmental health advocacy.

7. United States Environmental Protection Agency, EPA Pesticide Fact Sheets. *Protecting Children From Pesticides.* January 2002. Available at *http://www.epa.gov/pesticides/factsheets/kidpesticide.htm.* The EPA reports that "children are at a greater risk for some pesticides for a number of reasons. Children's internal organs are still developing and maturing and their enzymatic, metabolic, and immune systems may provide less natural protection than those of an adult. There are 'critical periods' in human development when exposure to a toxin can permanently alter the way an individual's biological system operates . . . Children's behaviors, such as playing on the floor or on the lawn where pesticides are commonly applied, or putting objects in their mouths, increase their chances of exposure to pesticides. Adverse effects of pesticide exposure range from mild symptoms of dizziness and nausea to serious, long-term neurological, developmental and reproductive disorders."

8. Carson, *Silent Spring.*

9. Rachel Carson to Dorothy Freeman, January 23, 1962, as quoted in Brooks, *The House of Life; Rachel Carson at Work.*

10. Rachel Carson to Lois Crisler, February 8, 1962, as quoted in Brooks, *The House of Life; Rachel Carson at Work.*

11. The US Fish and Wildlife Service reported that "homeowners use up

to 10 times more chemical pesticides per acre on their lawns than farmers use on crops." As quoted in Wargo, *Risks From Lawn Care Pesticides.*
12. *Albany Times Union,* March 2, 2006, page 1.

Chapter 11

The Green Fuse

Amsterdam, New York
2006

My neighbor who lives around the corner is a big man, with red hair and broad shoulders, but he's an affable sort. He waves as I stroll past his yard on a summer morning; it's early but the sun is already hot, and he is red-faced and sweating from his lawn work. "Take a break," I suggest.

"Gotta get this done," he says, and proudly shows me a new tool, the Weed Hound, a wickedly sharp-toothed instrument for digging up dandelion roots. The lawn behind him is smooth and green, the grass blades short and bristly as a Marine's crew-cut. "Gotta keep after it, otherwise the dandelions take over," he says, glancing down at the stubble. "I keep the grass short, don't want to give the little devils anywhere to hide."

He takes another whack at the earth with the sharp tool. I consider giving him a brief talk on vegetative reproduction, and helpfully pointing out that he's probably assisting the dandelion in its attempt to colonize his yard, but I reconsider; I'm a little afraid he'll set the Weed Hound on me. "You could eat them, you know," I point out hesitantly.

"I suppose," he says, without enthusiasm. "Now if we could just convince people dandelions are aphrodisiacs, like rhino horns, the problem would be solved." He plunges the blade into the soil again, then shakes his head. "Too bad there's no use for the damned things."

～

My hope is that after reading this book, you'll greet the advent of spring dandelions on your lawn with cries of joy. But I suppose it's just possible that, even after ten chapters, you still would prefer a dandelion-free yard lawn. So . . . how can we get rid of these classic uninvited guests?

It would seem that the solution to dandelions in the lawn is to mow the lawn more often, cut it really short, get all those jagged leaves and lanky stems out of the way, to give the grass a chance to grow. This, however, is not the case. It's an easy mistake to make, somewhat like driving faster when you're running out of gas, so as to get to the gas station sooner. Intuitively, it seems to make sense, but of course the paradox is that the more speed, the sooner you end up stuck by the side of the road. It seems illogical to slow down in order to make haste, or to stop cutting a plant in order to get rid of it. But counter-intuitive though it may be, the way to get rid of dandelions is to mow less.

Remember what dandelions like: sun, sun, and more sun. Seral species, remember? Shade is death. A dandelion would rather struggle up through a crack in sun-baked cement than do battle with shade. All the organic lawn manuals and websites that I consulted agree: raise the blade on your mower to the highest setting and let the grass grow 3 to 4 inches long. Researchers at the University of Maryland claim that mowing cool-season turfgrasses to 3 inches high works as well or better than herbicides to discourage crabgrass and other weeds.[1]

Again, it seems intuitive to cut the grass really short, especially if you hate yard work, so that more time will elapse until the next time you have to cut it. But it doesn't work that way. Plants have a built-in alarm system, an automatic response to disaster. When disaster (read: mower) strikes, the plant abandons its casual rate of growth and the slow process of building up nutrition in the roots before investing energy in creating seeds. It's wartime now, destruction is imminent! The plant invests all its efforts in a do-or-die sprint to get its genes into the gene pool. After a mowing, grass grows faster than ever, depleting the reserves in the root, squandering the hoarded energy in a desperate spurt. Dandelions do the same, rapidly sending up new leaves and hastily re-flowering. There's no time wasted on growing a new long stalk. This is why, after you mow a field of dandelions, they all seem to be back the next day, bright yellow flower heads defiantly sprouting out of half-inch-long stems.

I began to wonder how many folks out there were actually following the lawn manuals' wise advice about raising the mower blade. Deciding to follow Thoreau's practical example, I got out the ruler, and started crawling around in the grass. I measured the length of individual blades of grass in the lawns of half-a-dozen suburban yards, three school playgrounds, one park, and a shopping mall that had narrow strips of grass bordered by concrete curbs. The average grass blade length was 1.6 inches. Dandelion heaven.

Okay, so you need taller grass. What else? Well, a dedicated landscaper with a small home lawn can win the war battle by battle. Pour vinegar on each and every dandelion, and reapply several times if they re-sprout. Or you can buy various tools like the Weed Hound, and try to avoid breaking the roots into bits that will clone. Or you can forget the

Weed Hound and really get serious — you can purchase the Weed Dragon, a fire-breathing mini-flame-thrower that scorches individual dandelions out of existence.

But I hadn't realized how determined dandelion-haters can be till I discovered the Ultimate Weapon: the Dandelion Terminator. It's a sort of giant hand drill that you hold like a gun — it's really worthy of Arnold Schwartzenegger. You take aim at point-blank range and pull the trigger. The Terminator whines and roars, and drills that dandelion root into smithereens.

It's okay, though, I don't mind. These lethal weapons are harmless, unless they set your house on fire, or accidentally drill out your eye. These tools will kill no songbirds, poison no butterflies; they're environmentally benign, unlike the innocuous-looking bag of herbicide from the garden store. The Dragon and the Hound and the Terminator will help keep your yard weed-free, but they'll no more eradicate dandelions than stepping on a few ants will make ants go extinct. Dandelions are here to stay.

～

Here's a question to ponder. What would you consider to be the most successful species of life on earth?

"Humans!" all we humans shout. Well, of course, it's taken for granted. But the issue is, at the very least, debatable.

I suppose it depends on how you define success. By sheer numbers of individuals? Ants beat us every time. Ability to reproduce rapidly? Norway rats are pretty good at that. Happiness? Are humans happier than, say, golden retrievers? Hard to say.

You could claim, perhaps, that a species that has spread itself all over the globe is successful. Humans have done that pretty thoroughly. Another mark of success might be the ability to enslave another species, to make that species work for you, aid in your global spread, even fight off other organisms that threaten you. By this criterion, you could argue that wheat is the most successful species on the planet, and that wheat has become that way by enslaving another species: *Homo sapiens*. Humans have not only spread wheat around the globe, but aggressively seek to destroy all other species that would encroach on the successful growth and spread of wheat.[2] Agriculture, like pollination, may be just another one of plants' many schemes to put the Animal Kingdom to work.[3] It's an interesting point of view.

In the case of wheat, or rice, or any crop plant, humans are indeed rather like slaves working to support a master. But we have entered the contract willingly, with our eyes open; one could argue that it isn't slavery at all, but a partnership, in which we receive the great benefit of the plants' bountiful nutrition in return for our services. But, recently at least, dandelions haven't had the benefit of our assistance — quite the opposite — yet they still thrive.

One criterion for global success is the ability to adapt to changing conditions. And surely the key to dandelions' world conquest is their adaptability. Take a look at the long, sad list of plant and animal species that are threatened or endangered worldwide. It's filled with organisms that are less than adaptable. Many are tied to certain very specific environments, like a lovely wildflower called sandplain gerardia, or saltmarsh bulrush, or tundra dwarf birch. In other words, they're picky about where they live. Or they're picky about their food, like the dainty Karner blue butterfly, that as a caterpillar will eat absolutely nothing but the leaves of a single species of a rather uncommon plant, blue lupine. Other species are picky about how they reproduce, perhaps limited to a single species of pollinator or a single disperser of their seed. But being picky is the surest way to get to the top of the endangered species list.

Dandelions are like the determined survivors of a shipwreck or plane crash that will eat anything, use anything, do *anything,* to stay alive. There are a few other such hardy creatures, ubiquitous animals like raccoons, gulls, coyotes, and crows, to name a few — organisms that have adapted as the face of their world changed, opportunists that find a way to thrive in widely varying environments, in close proximity to humans.

Dandelions carry this to an amazing extreme. For the vast majority of flowering plants, pollination is crucial. No pollination, no seed. But dandelions don't need the birds and the bees to carry pollen for them. Dandelions can make seed on their own — at least some individuals can, depending on the number of chromosomes that they have. Like a weird sort of botanical virgin birth, some dandelions can produce fertile seed without being pollinated at all, by themselves or by another plant. This is a process called apomixis, a peculiar habit shared by only a few other types of plants — approximately 400 species out of the vast array of hundreds of thousands of plant species in the world — and one that frequently leaves botanists scratching their heads.

The seeds produced apomictically are genetically identical to the parent plant. Lots of plants, from potatoes to tulips, can clone their roots, tubers or bulbs, but very few can clone their seeds. Each of these clonal lineages reproduces itself as a distinct "microspecies." The tendency to apomixis makes dandelions ridiculously hard to classify, since each strain of plants could arguably be considered a distinct species, and botanists grow heated to this day over exactly how many species of *Taraxacum* there are — some say thousands. The accepted number of dandelion species varies wildly, depending on whether one is a "lumper," lumping a group of similar plants together into the same species despite minor differences, or a "splitter," splitting a group of similar plants into more than one species on the basis of minor differences.

This tendency to apomixis gives the dandelion a stunning advantage when it comes to reproduction — no matter if rain or snow keeps pollinators away, no matter if disease wipes out all the honeybees, no matter if there isn't another dandelion for a hundred miles. An apomictic dandelion is utterly self-sufficient. Unlike almost every other plant you see in your yard, the dandelion does not depend on another organism, of its own species or another, to produce seed. The sepals close around the flower head, and bam! — seeds magically appear. An added benefit to apomixis is that while a plant that clones a root generally passes on any disease from the mother plant to the clones, pathogens are rarely transmitted through seed. It's impossible to say for sure what percentage of dandelions are apomictic, but botanists estimate that most of the dandelions in the New World are able to reproduce through apomixis — it's the rule rather than the exception.

This is an adaptation that is ideal for urban environments. It explains why dandelions can pop up even in the heart of the city, in places where there isn't a honeybee in sight. Dandelions may be evolving in the direction of not having flowers at all.[4]

Like Linnaeus, Culpeper, and Thoreau, scientists are once again studying the humble dandelion. This time they are examining this weird trait of apomixis, and some are asking tantalizing questions: What if this superb adaptation for survival could be triggered in other plants? Plants like wheat, or corn, or rice? None of the global cash crops are apomictic by nature, but scientists are attempting to induce apomixis in food plants by genetic engineering.[5] This would enable the rapid development of hundreds of varieties of rice or wheat, each genetically suited to a particular micro-climate. A type of wheat that grew really well in a particular arid valley in a hungry Third World country could be made to produce seed apomictically, so that all the ensuing crops would have those precise traits that made the plant so successful in that one spot. Enthusiasts say that the study of apomixis in dandelions and other plants could revolutionize agriculture.

On the other hand, that's what they said about DDT. Then they discovered that the miracle pesticide had a few side-effects, like wiping out bald eagles. The possibilities for success are endless — and so are the possibilities for disaster. But whether genetic engineering thrills you or appalls you, dandelions are right now on the cutting edge of science — where they've always been.

In the twenty-first century it seems as if we could live, if we wanted to, totally removed from the natural world. We have apparently unlimited technology at our command — we can control nature, change it, design it. Or we can escape it; we can build walls, erect barriers, to keep us separated from the wild. Long ago, it must have seemed to the Pilgrims and pioneers as if they were on a little island of civilization in a threatening sea of wildness, and dandelions were part of their fence to keep the wilderness at bay. Nowadays, it seems as though the wild places are the fragile islands, surrounded by a rising tide of pavement.

But as we painfully discovered during Hurricane Katrina, no dam or levee can finally hold back the surge of nature. The poet Dylan Thomas described "the force that through the green fuse drives the flower," and he well might have had in mind the quiet explosion of a dandelion thrusting its way through a crack in the concrete. No weed-whacker or poison yet invented has so far exterminated those determined little flowers — they continue to follow us. As they always have, since time out of mind.

A warm spring day, and the morning sun is striking the columns of the ancient temple; the old marble glows the color of honey against the blue Mediterranean sky. On my fiftieth birthday, I gave myself a present, a two-week trip to Greece — the cradle of Western civilization, where it all began.

It's where things all began for the dandelion, too. This hot, sunny, dusty part of the world is where dandelions first evolved. Walking the olive groves near the ruined temples, I'm amazed at how many variations on the dandelion theme flourish here. It's nothing but sun and disturbed soil, their favorite habitat: Compositae abound. I see round yellow blossoms on head-high stems, like dandelions on stilts; a huge puff-ball the size of a grapefruit, a dandelion on steroids. There are tiny dandelion-like flowers the size of peas, trodden underfoot on the dusty track under the cypress trees.

And dandelions themselves, *Taraxacum officinale*, the familiar jagged-leaved, lion-toothed American lawn weeds, grow everywhere in Greece. Tourists these days can't go inside the Parthenon — you have to stand behind a barricade, and gaze from afar at the beautiful columns. But the rowdy dandelions jump the fence and spring out of cracks between the massive marble steps of the temple, as they no doubt did in the days when Socrates and Plato walked there. Dandelions poke out of crevices in the ancient bedrock below the Parthenon, too, comfortably at home as they have been since time out of mind, long before Neolithic humans first climbed the Acropolis. Hey, the dandelions were here first.

We rarely notice dandelions these days, except when we try to kill them. No more do we seek their health-giving benefits, or their rich nutrition. Instead we try our best, with venomous ingenuity, to exterminate them, but they spring back stronger than ever. Then we ignore them, but still they follow us, down the centuries; they stick to us as closely as a dog, as inevitable as a shadow. We may try to outrun them, but we never will; they are our footprints.

Notes to Chapter 11

1. *http://hgic.umd.edu*, the website of the University of Maryland, College of Agriculture and Natural Resources' Home and Garden Information Center.

2. Attenborough, David. *The Secret Life of Plants.* London, UK: BBC Books, 1995.

3. Pollan, *The Botany of Desire.*

4. Stokes, Donald, and Lillian Stokes. *A Guide to Enjoying Wildflowers.* Boston, MA: Little Brown and Co., 1984.

5. *http://www.biotech-monitor.nl*, the website of *The Biotechnology and Development Monitor,* a quarterly publication of the Network University in Amsterdam, The Netherlands.

Bibliography

Albany Times Union, March 2, 2006.

Angier, Bradford. *Field Guide to Edible Wild Plants.* Harrisburg, PA: Stackpole Books, 1974.

Attenborough, David. *The Secret Life of Plants.* London, UK: BBC Books, 1995.

Blunt, Wilfrid. *The Compleat Naturalist: A Life of Linnaeus.* New York, NY: Viking Press, 1971.

Botkin, B. A., ed. *A Treasury of New England Folklore.* New York, NY: Bonanza Books, 1947.

Bradbury, Ray. *Dandelion Wine.* New York, NY: Bantam Books, 1964.

Bradford, William. *Of Plymouth Plantation.* New York, NY: Random House, 1981. First published as *History of Plymouth Plantation* in 1856.

Bradley, Fern, ed. *Chemical-Free Yard and Garden.* Emmaus, PA: Rodale Press, 1991.

Briggs, Robin. *Witches and Neighbors.* New York, NY: Penguin Books, 1996.

Brooks, Paul. *The House of Life: Rachel Carson at Work.* Boston, MA: Houghton Mifflin, 1972.

Brown, Tom. *Guide to Wild Edible and Medicinal Plants.* New York, NY: Berkley Books, 1985.

Buckland, Raymond. *Witchcraft from the Inside.* Saint Paul, MN: Llewellyn Publications, 1971.

Burdick, Alan. *Out of Eden: An Odyssey of Ecological Invasion.* New York, NY: Farrar, Straus and Giroux, 2005.

Burns, Deborah. *Shaker Cities of Peace, Love and Union.* Hanover, NH: University Press of New England, 1992.

Carson, Rachel. *Silent Spring.* Boston, MA: Houghton Mifflin, 1962.

Culpeper, Nicholas. *Culpeper's Complete Herbal,* Herefordshire, UK: Wordworth Editions, 1995. First published in 1652. Also available at *http://www.med.yale.edu/library/historical/culpeper/intro.htm.*

Drake, Samuel. *Annals of Witchcraft in New England.* New York, NY: Benjamin Blom, 1967. First published in 1869.

Dutton, Joan Parry. *Plants of Colonial Williamsburg.* Williamsburg, VA: The Colonial Williamsburg Foundation, 1979.

Elias, Thomas, and Peter Dykeman. *Field Guide to North American Edible Wild Plants.* New York, NY: Outdoor Life Books, 1982.

Elliott, Douglas B. *Roots: An Underground Botany and Foragers' Guide.* Old Greenwich, CT: The Chatham Press, 1976.

Erichsen-Brown, Charlotte. *Medicinal and Other Uses of North American Plants.* New York, NY: Dover Publications, Inc., 1989. First published as *Use of Plants for the Past Five Hundred Years* in 1979.

Fara, Patricia. *Sex, Botany and Empire,* New York, NY: Columbia University Press, 2004.

Gerard, John. *The Herball; or General Historie of Plantes.* New York, NY: Dover Publications, 1975. First published in 1597 and revised in 1633.

Gibbons, Euell. *Stalking the Healthful Herbs.* New York, NY: David McKay, Co., Inc., 1971.

———— *Stalking the Wild Asparagus.* Brattleboro, VT: Alan C. Hood and Co., Inc., 1962.

Grieve, Maud. *A Modern Herbal.* New York, NY: Dover Publications, Inc., 1971. First published in 1931.

Huntford, Roland. *The Last Place on Earth.* New York, NY: Random House, 1983.

Josselyn, John. *New-England's Rarities Discovered.* Boston, MA: Massachusetts Historical Society, 1972. First published in 1672.

Judge, Michael. *The Dance of Time: The Origins of the Calendar.* New York, NY: Arcade Publications, 2004.

Karlson, Carol. *The Devil in the Shape of a Woman; Witchcraft in Colonial New England.* New York, NY: Penguin Books, 1987.

Kirk, Ruth. *Snow.* Seattle, WA: University of Washington Press, 1980.

Köhler, F. E. *Köhler's Medizinal-Pflanzen in naturgetreuen Abbildungen mit kurz erläuterndem Texte...* Gera, Germany: Gera-Untermhaus, 1883.

Kurlansky, Mark. *Cod: A Biography of the Fish That Changed the World.* New York, NY: Penguin Putnam, 1997.

Lawn Pesticides, An Unacceptable Risk, a Publication of Grassroots Environmental Education, 2003.

Leighton, Ann. *American Gardens in the Eighteenth Century: For Use or For Delight.* Boston, MA: Houghton Mifflin, 1976.

Leopold, Aldo. *A Sand County Almanac,* New York, NY: Oxford University Press, 1968. First published in 1949.

Manning, Richard. *Grasslands: The History, Biology, Politics, and Promise of the American Prairie.* New York, NY: Penguin Books, 1997.

Martin, Alexander C., Arnold L. Nelson, and Herbert S. Zim. *American Wildlife and Plants: A Guide to Wildlife Food Habits.* New York, NY: Dover Publications, Inc., 1951.

Millspaugh, Charles F. *American Medicinal Plants.* New York, NY: Dover Publications, Inc., 1974. First published in 1892.

Morse, Roger A. *The Complete Guide to Beekeeping.* New York, NY: E. P. Dutton, 1972.

New York Flora Association Newsletter, Vol 13, No. 1, March, 2002.

Peterson, Roger Tory, and Margaret McKenny. *A Field Guide to Wildflowers: Northeastern and North-Central North America.* New York, NY: Houghton Mifflin, 1968.

Pollan, Michael. *The Botany of Desire.* New York, NY: Random House, 2002.

Rachel Carson Council News, No. 91, Spring, 1999.

Richardson, Joan. *Wild Edible Plants of New England.* Yarmouth, ME: DeLorme Publishing Co., 1981.

Richardson, Robert D. *Henry David Thoreau: A Life of the Mind.* Berkeley, CA: University of California Press, 1986.

Robertson, Laurel, Carol Flinders, and Brian Ruppenthal. *The New Laurel's Kitchen.* Berkeley, CA: Ten Speed Press, 1986.

Sanders, Jack. *The Secrets of Wildflowers.* Guilford, CT: The Globe Pequot Press, 2003.

Selsam, Millicent E. *The Amazing Dandelion.* New York, NY: William Morrow and Co., 1977.

Shakespeare, William. *Macbeth.* Edited by Robert S. Miola. New York, NY: W. W. Norton and Co., 2004.

Silverman, Maida, *A City Herbal.* New York, NY: Alfred A. Knopf, 1977.

Solecki, Ralph S. *Shanidar: The First Flower People.* New York, NY: Alfred A. Knopf, 1971.

Stein, Sara. *Noah's Garden: Restoring the Ecology of our Own Backyards.* New York, NY: Houghton Mifflin, 1993.

Steinberg, Ted. *American Green: The Obsessive Quest for the Perfect Lawn.* New York, NY: W. W. Norton and Co., 2006.

Stokes, Donald, and Lillian Stokes. *A Guide to Enjoying Wildflowers.* Boston, MA: Little Brown and Co., 1984.

Stone, Irwin. *The Healing Factor: Vitamin C against Disease.* New York, NY: Grosset and Dunlap, 1972.

Sturtevant, E. Lewis. *Edible Plants of the World.* New York, NY: Dover Publications, 1972. First published as *Sturtevant's Notes on Edible Plants* in 1919.

Thoreau, Henry David. *Faith in a Seed: The Dispersion of Seeds and Other Late Natural History Writings.* Edited by Bradley P. Dean. Washington, DC: Island Press, 1993.

———. *Wild Fruits.* Edited by Bradley P. Dean. New York, NY: W. W. Norton and Co., 2002.

————. *Walden.* Philadelphia, PA: Running Press, 1987. First published in 1854.

Tilford, Gregory. *From Earth to Herbalist.* Missoula, MT: Mountain Press Publishing Co., 1998.

Tilford, Gregory, and Mary Wulff-Tilford. *Herbs for Pets.* Irvine, CA: BowTie Press, 1999.

Travers, Carolyn Freeman, ed. *The Thanksgiving Primer.* Plymouth, MA: Plimouth Plantation, Inc., 1987.

Turner, William. *A New Herball.* Cambridge, UK: The Cambridge University Press, 1989. First published in three parts in 1551, 1562, and 1568.

Walker, Winifred. *All the Plants of the Bible.* Garden City, NY: Doubleday and Co., Inc., 1979.

Wargo, John. *Risks from Lawn Care Pesticides.* North Haven, CT: Environment and Human Health, Inc., 2003.

Weed, Susun, *Healing Wise.* Woodstock, NY: Ash Tree Publishing, 1989.

Wheelwright, Edith Grey. *Medicinal Plants and Their History.* New York, NY: Dover Publications, Inc., 1974. First published as *The Physick Garden: Medicinal Plants and Their History* in 1935.

Willison, George F. *Saints and Strangers; Being the Lives of the Pilgrim Fathers, & Their Families, With Their Friends and Foes.* Orleans, MA: Parnassus Imprints, Inc., 1945.

Wolverton, B. C. *How to Grow Fresh Air.* New York, NY: Penguin Books, 1996.